Farmers of Five Continents

Farmers
of
Five
Continents

Don Paarlberg

University of Nebraska Press

Lincoln and London

The paper in this book
meets the guidelines for
permanence and durability
of the Committee on
Production Guidelines for
Book Longevity of the
Council on Library Resources.

Library of Congress
Cataloging in Publication Data

Paarlberg, Don, 1911–
Farmers of five continents
1. Farmers. 2. Agriculture.
I. Title.   S439.P2   1984
338.1   83-17090
ISBN 0-8032-3670-0

Cover photo courtesy
of Donald G. Hanway,
Professor of Agronomy
at the University of
Nebraska–Lincoln
and former coordinator
of the Morrocco
Dryland Farming Project.

Endsheet photo by
Peyton Johnson,
courtesy of the
Food and Agriculture
Organization of the
United Nations,
Rome, Italy.

Contents

Preface

Food is man's most ancient problem. An old Byzantine proverb has it thus:

> He who has bread has many problems;
> He who lacks bread has only one problem.

We in the United States, the fortunate few, have abundant food and so can afford the luxury of other problems.

In much of Asia, Africa, and Latin America mankind lives almost literally from hand to mouth. The carry-over of food from one crop year to the next is never more than a small share of the coming year's needs. A failed crop means at least the threat and possibly the fact of famine.

Most notable among the famine-fighters are the farmers, who man the front-line trenches. Worldwide, farm people are more numerous than all other people combined. They are direct, straightforward, and immensely practical. These famine-fighters, like many other battlers, are not exclusively occupied with the enemy. They are whole persons, alive and active. I find them more interesting than Egyptian pharaohs and Chinese emperors, who are all dead.

In 1982 my wife Eva and I interviewed farmers in seven countries. This was a private undertaking, unsponsored and at our own expense. Earlier I or the two of us had visited farmers in four other countries. On each of these farms we asked such questions as: "What farming system is followed?" "What are the practices and what is the performance?" "What are the aspirations and the problems?" "What are your staple foods?" "Tell us about your family." I have tried to distill the essence of the interviews in the separate sections of this small book.

The farmers were chosen in a variety of ways. Some were selected

with the help of friends. Our Purdue University connections were
helpful. Some were picked with the assistance of the government of
the host country. The United States Foreign Agricultural Service
helped, as did the United States Agency for International Development. The International Association of Agricultural Economists as-
sisted in choosing three of the farmers. Several were chosen purely by
chance. All were authentic. They constitute a series of case studies;
they may or may not be representative of a larger group. On the whole
their farms are above the average in size and performance.

The farmers portrayed here are very diverse. For some of them
famine is an ever-present threat. For others it is a vivid memory. For
most of them it was an experience of the ancestors, long forgotten.

Some of the farm operations are large, and some are small. Among
those reported are farmers young and old, traditional and modern,
producers of crops and of livestock, entrepreneurial and centrally
directed. In formal education they vary from zero to college graduates.
There are in this book farmers from the Northern and Southern
hemispheres, from tropical and temperate zones, from East and West.
They come from each of the five continents and from the islands of
the sea. They come from countries in which the dominant belief,
philosophy, or nonbelief may be Christian, Islamic, Hindu, Buddhist, Confucian, or atheistic.

While the differences are striking, there are similarities. All are
famine-fighters, whether or not knowingly enrolled for battle. Each is
dependent on the bounty or parsimony of nature. Each contends with
the weeds and the weather. Seedtime means hope for each, and
harvest brings satisfaction or disappointment, as the case may be.
Farmers are brought into continual awareness of the wonder of birth,
growth, and death. These similar experiences mean a kinship of countrymen, however diverse their backgrounds.

The book is more like a series of solo voices than like a choir. Farm
people can perhaps find in it some harmonious chord. For nonfarmers it may awaken ancestral memories that long have slumbered.
Farming is the heritage of us all. If one does not farm, his predecessors
did, one, two, five, or ten generations past. Some residue of this
tradition lingers on.

It has been a joy to meet these farmers, to visit with them, and to
fashion their stories into this book.

*   *   *

In what order should these farmers be presented? Chronologically, as interviewed? By geographical grouping? In order of size? According to the state of agricultural technology? By degree of entrepreneurship? With respect to presumed reader interest? Every counselor had a different idea, and every alternative posed a problem. I finally decided to list them by countries, alphabetically, clearly identified in the contents. Thus the reader may proceed however he wishes. The sections are written so they may be read in any order. My advice is to start with the one that promises to be of greatest interest and so read on.

Bali

Fatal Impact?

Denpasar, September 3, 1982. Bali is an island gem lying just east of Java, midway in the Indonesian nation, a volcanic upthrust, dominated by Gunung Agung, twelve thousand feet high, still active, occasionally belching out ashes, rocks, and molten lava. Bali is about the size of Delaware. Rainfall is heavy, but not adequate for the intensive cropping system. The island is heavily populated, with 350 people per square mile, five times as dense as in the United States. The inhabitants are genetically akin to the other people of this watery world: the Malaysians, the Filipinos, and their Indonesian brothers. But their religion and their culture are Hindu.

The people of Bali have gotten it all together: the biological and the social, farm and nonfarm, art and livelihood, sacred and secular, good and evil. To the Balinese, good and bad are identifiable but inseparable. Both are present in every act and every event; neither is ultimately triumphant.

We Westerners differentiate the various parts of living: the arts from the sciences, the various professional disciplines from one another, the sacred from the secular, the myth from the fact, and the good from the bad.

We pull ourselves up by the roots to see how we are doing. The Balinese leave themselves rooted. We tear ourselves apart and place on the psychiatrists the responsibility for putting us back together. The Balinese integrate themselves, leaving a smaller task for outsiders.

The institutional arrangement with which the Balinese farmers achieve integration is a cooperative irrigation system called the *subak*. The word *subak*, of great age and Indian origin, comes from the Sanskrit words meaning "together" and "water." The system is in fact at least one thousand years old. There are in the island of Bali more than a thousand of these irrigation cooperatives, and they include altogether about a quarter of a million acres, averaging almost 200 acres for each cooperative. The subak we are going to visit is named Umedesa. It has in it 229 farmers and totals 250 acres, a little more than an acre per farm. American farms average 430 acres each.

The subak Umedesa lies outside the city of Denpasar, amid a cluster of the rural villages that are characteristic of the island. The volcanic soil is cut deep by valleys that drain south toward the sea. Valley floors are wide or narrow, according to the whim of the descending waters. Hillsides are terraced into five, ten, or twenty levels, or even as many as fifty. Some of the terraces are only a few feet wide. They were laid out centuries ago, for hand tillage, when there was no thought that riceland would ever be plowed with bullocks or tractors. On many fields the terrain now rules out anything but hand labor. The terraces are perfectly conformed to the landscape. Rice is almost everywhere. When rice is in midgrowth, it is bright green, almost iridescent. Thin streams of irrigation water, sparkling in the sun and carefully controlled, spill from higher to lower levels. One of the hillsides, seen from across the valley, looks like a crown encrusted with emeralds, interspersed with silver. Pictures and paintings of this island, which I had thought imaginary, are indeed true.

Rice is in all stages of growth. Some is being harvested. It is cut with a curved sickle, a recent advance over the *ani-ani*, a hand-held knife. Some degree of animism survives in this part of the world. The grain must be removed from the stalk with respect, lest the spirit of the rice be offended. The rice is threshed by grasping a double handful of rice stalks and flailing the heads against a board. Winnowing follows and then drying in the sun, spread thin on mat, hard ground, or concrete floor. There are scarecrows and white pennants to frighten off the birds. The rice straw is burned after threshing. There are no bullocks to eat it, and it will not decompose fast enough if turned underground. So it is burned; there is a haze of smoke in the air. Straw-burning is probably one of the few faults the environmentalist would lay against this farming system.

We come to the subak. There is the headquarters building, a temple, a rice-drying floor, and a fish pond. We meet the head man, Ineugah Gedoa, one of the farmer members. He is forty years old, a rugged handsome man with close-cropped hair and a short black beard, the father of two sons and a daughter. He wears a tan bush jacket in the pocket of which is a ball-point pen, testimony to his six years of formal education. He was elected three years ago for a five-year term. He wears the wraparound skirt called a sarong, green with a yellow sash. He is barefooted. His language is Indonesian. Interpretation is by Raka Djaja, an agricultural extension officer. We talk at the subak headquarters, a pleasant open place where the members hold their meetings. These occur at intervals of about thirty-five days, the length of the Balinese month. The particular day of the meeting is chosen by the spiritual leader, who discerns the auspicious time. On the wall is a map showing the irrigation scheme and the landholdings. There are three subsystems, each organized under a lieutenant. Mr. Gedoa is the overall chief. The questions begin.

"Where do you get water?"

"There is a big dam six miles up the river," he says. "It has been there many years. It was renovated about ten years ago."

"How do you regulate water use?"

"It regulates itself. My farm is at the lower end of the system. If I do not get water, there is trouble. We make our own rules and regulations. It is the *awig-awig* system."

"How do you discipline a member if he gets out of line?"

"First we warn him. If he does not change, he gets no water."

"Have you had to shut anyone off?"

He can't remember the last time anyone was shut off.

"What are the responsibilities of the members?"

"Keep the ditches and canals in repair," he says. "Abide by group decisions as to variety of rice and time of planting. Help build new facilities. Participate in religious ceremonies. Attend meetings and elect officers. Make contributions for improvements."

"What is the rice yield on this subak?"

We learn that there are two crops of rice a year, each averaging more than three tons per acre. In addition there is a crop of maize or sorghum or peanuts that produces about one ton per acre. This is a total of seven tons to the acre. Such production is possible with multiple cropping, fertile soil, year-round growing temperatures, ample water, abundance of labor, and good management. We have never heard of such yields, let alone seen them. The average American yield of corn, our most productive crop, is less than half as much, about three tons per acre.

"Tell us about the rice."

The gist of his long response is as follows: It is IR 50, bred by the International Rice Research Institute in the Philippines. It is a short-season, rapid-growing variety, ninety-five days from planting to harvest. By mutual agreement, all members grow the same variety. They phase the planting so as to make for even use of water and to minimize peak labor loads. The farmers trade labor. Rice is fertilized according to recommendations of the extension service. Chemicals are used to control pests and diseases. Tillage is by hand with the broad-bladed mattock found throughout the Third World. There is no animal draft power. Occasionally a farmer is short on time in getting ready for the next crop. If his field is of a size and location to permit and if he has the money, he may hire a tractor. There are a few small Japanese tractors in the village, available for custom work.

Such a combination of the traditional and the modern!

Farmers use rice for the family food supply and sell the balance to rice merchants and rice mills in the village.

"What part of the crop is sold?"

"If a farmer has half an acre of land and has a family of six people, he can sell about half his rice," he says.

We Americans think of agriculture as producing crops for sale, with which income to buy food and other things. We need to remind ourselves that in most of the world agriculture consists of producing crops for food use at home, selling only the small excess above family needs.

"Does the subak have responsibility for schools and health and law abiding?"

"No," he tells us. "These things we leave to the civil authorities. We are concerned with three things: the land, the people, and religion."

"Please tell us about these things."

To summarize a long discussion: Land is privately owned. The various holdings are similar in size. Most land is operated by the owner, but some is rented. The land is divided among the sons when the father dies. The daughters get no share; their circumstances are determined by the inheritance of the men they marry.

The people all belong to the subak, whose natural boundaries are fixed. They have long worked and worshipped together. Women have a full share of the labor, though none of the inheritance. Farms are so small that they don't require full-time labor of all the family members.

Some work part-time at construction or other tasks in the village. We learn that some of the wood carving we see in the shops is done by these farmers. In the evening, with a group of tourists, we attend a traditional dance in the village. One hundred men sing and chant, a human organ, doing the *kecak*, or monkey dance. We learn that some of the dancers work in the rice fields during the day. The people are brought up in the setting of respect for the artistic tradition. The children at play assume the acts and postures of the dance.

The Hindu religion is clearly a binding force. Oral tradition and the dance are based on the Hindu classics. The priests are consulted on matters of importance. Life centers on religious festivals and ceremonies. We see processions of brightly dressed people, the women carrying on their heads great baskets of rice and fruits. After presenting this food to the gods and the spirits of their ancestors at the local shrine, it is borne back home again, leaving only a token morsel. This deportment should not be puzzling to a Westerner who lifts his glass of wine in a toast, says "Here's to good old Joe!" and drinks it himself.

"What do the people eat?"

This baffles Mr. Gedoa, who seemingly has fixed his mind so completely on rice that the question seems ridiculous. "Rice," he says. Then, as if pulling the information from some hidden depth, he slowly adds: "Fruit, from the trees in the dooryard. Fish from the pond. Vegetables from the field boundary. Eggs and meat from the ducks that glean the fields after harvest. Sometimes cassava. But mostly rice."

We learn that to the east, on the island of Timor, the food supply has not been adequate and there has been famine. During recent history Bali has escaped.

"How well do the people live?"

"We have electricity. More than half of us have television sets; we get programs from the new station at Denpasar."

This astounds us.

We note that Mr. Gedoa has a small Japanese motorbike.

We learn that the extension service uses this subak for demonstration meetings, teaching better methods to farmers from other cooperatives.

"Does the subak system exist elsewhere in Indonesia besides in Bali?"

The extension man responds. The government has tried to plant it elsewhere, but, lacking tradition and religious motivation, it has not taken hold.

"What is the rate of population growth in Bali?"

Mr. Djaja makes answer. "During the seventies the population grew 1.7 percent annually."

We note that at this rate the population will double in about forty years. We know from our base book that there is a family planning program, but it is weak. The very tradition that gives the culture its strength is an obstacle to a successful family-planning effort.

In ancient days, when the tropical islands became overpopulated, the chief designated certain young men and women to explore for a new island. The people were named, the boat built and provisioned, the prayers said, and off they went toward the horizon. They might find an island, or they might not. In any case they did not come back. Pride and peer pressure prohibited.

We learned there is a transmigration program in Bali. The government resettles those who, on a voluntary basis, are willing to leave overcrowded Java and Bali for the thinly populated areas of Sumatra, Kalimantan, Sulawesi, and West Irian. But few elect to go; friends, family, tradition, and social reinforcement surround them in the home island, poor though they be. During the six-year period from 1966 through 1971, more people migrated to Java and Bali from the other islands than the government was able to move in the opposite (desired) direction.

In Bali the scale that equates food and people is precariously balanced. Almost all available land is in use. Hardly a drop of water flows unused to the sea. The cropping pattern is highly intensive. The diet is about as heavily dependent on the high caloric product, rice, as good nutrition will condone. Industrialization and exports, making possible exchange earnings and the importation of food, proceed slowly in a society with a strong rural culture. Off-farm jobs thus are scarce. Population continues to grow. In these circumstances hunger looms as a possibility.

"Any more questions?" asks Mr. Djaja. We ask no more.

But as we leave the subak, a brooding unasked question occupies my mind, pushing other thoughts aside.

What is the prospect for this traditional society in a modernizing world? The Balinese have a remarkable record. Their culture survived intrusion of the Chinese, the Portuguese, the Dutch, and the Japanese. It survived the chaotic years of insurgency that followed the gaining of independence in 1945. These people successfully resisted the surge of Islam, which became implanted in virtually all the other islands of what is now Indonesia. Can this established culture meet the triple threat of technology, television, and tourism? Will this traditional society be subjected to the fatal impact experienced by other integrated agricultural communities when descended on by the modern world? If this time-honored culture were to be replaced by a modern society, what would be the gains and what the losses?

A bit of advice, once spoken by a political friend, comes to mind: "Don't worry about answers to questions that aren't being asked." This remembered counsel relieves the felt need for an answer. But it doesn't dismiss the question.

Brazil

The Laboratory

Brazil, August–September, 1973. We are attending the Fifteenth International Conference of Agricultural Economists at São Paulo. After a week of lectures a number of us lay a plan to go out into the country to see the farms and the farmers firsthand.

Brazil is a good laboratory for such a venture. The country has most of the different forms of agriculture, thus providing an opportunity for contrast and comparison. Agricultural development is under way, employing a variety of systems.

The dominant agricultural institution in Brazil is the *fazenda*, a large-scale farming unit with land, labor, and management supplied separately. So we begin there.

The fazenda we visit is named Monte d'Este (Mountain of the East.) It is located some distance from the city of Campinas, about one hundred miles inland from São Paulo. On the way to it we drive through the heart of Brazil's agricultural region; it has long been farmed and is highly productive. We are on a plateau tilted toward the west, rolling, with long easy slopes. We are close to the Tropic of Capricorn, about as far south of the equator as Miami is north of it. August in the Southern Hemisphere is late winter or early spring. Temperature is pleasant, and weather is fine. Fields are plowed, ready for planting. We notice some erosion. Rainfall is sufficient for good crop production. Soil is reddish yellow, typical for a warm humid country. We pass much coffee, coming into bloom. The small white blossoms are fragrant; there is no coffee-ness to their smell. The area also produces cotton, corn, sugar cane, soybeans, peanuts, castor beans, potatoes, and rice, as well as many vegetables. Among the fruits are warm-weather crops like citrus, banana, and pineapple.

There are temperate climate fruits such as grapes, peaches, and apples. We see manioc, the tall stalky plant with a starchy root, used for food throughout the warm countries of the world. Otherwise known as mandioca and cassava, it is the plant from which tapioca is made.

Fields are large. Farmsteads are far apart and consist of groupings of tile-roofed buildings. There are clumps of trees—pine and eucalyptus, and occasional palm.

Fazenda Monte d'Este is a very large operation, home of one hundred families. During the peak of the coffee harvest, thousands of part-time workers are employed. There are about twenty-five hundred acres in the fazenda. Coffee is the main enterprise. Certified cottonseed is produced, as well as hybrid seed corn. Grapes, pineapple, and citrus are grown. There is a dairy herd of three hundred Holstein cattle. All this we learn from a handout supplied by a Brazilian colleague.

We approach by a lane that winds through a eucalyptus grove, past the small chapel that serves the fazenda. We come to a broad open area in what looks like the center of a village but is actually the headquarters of the farm. There is a small lake and a playground.

In the United States we would be greeted by the owner and given a brief explanation, after which there would be questions. Here the hosting is more casual. We are given no financial data; we are simply invited to look about.

Some of us investigate the coffee complex. It is a large two-story building, with four huge brick drying-floors. Processing involves fermenting the coffee berries so that the pulp can be removed, then taking off the husk that encloses the paired beans, after which the beans are roasted. All is now deserted, it being the off season.

We are invited to visit the mansion. It develops that the owner is Japanese and lives in Tokyo. Many of Brazil's largest farms are owned by individuals who live in the city, often at a distance. But few live in other countries. The mansion, which is unoccupied, is kept in good condition. It is a beautiful home of masonry with tile roof and floors. Furniture is dark, angular, and heavy, in the Portuguese style. Windows are shuttered. Colors are pastel. We feel a bit ill at ease, wandering through someone else's house.

Grouped about, at a respectful distance from the mansion, are the homes of the families who live and work here. These are small, of masonry, most of them in a sheltering grove. Each has a small garden behind a white picket fence. We see a few of the women and some children but almost no men.

There is a large building for the dairy herd, but we are admonished to stay away so as not to run the risk of introducing some disease.

We see farm equipment such as tractors, wagons, and the like. Obviously the fazenda has good working capital.

We wander up the dirt lane and see a good citrus grove. Next is the coffee, a very large area. The coffee, we are told, is threatened by rust, a fungus disease. Coffee breeders of the Agricultural Institute at Campinas are trying to develop immune varieties.

One of the Brazilians explains that the engineers are working on a coffee-harvesting machine, based on the principle of the American blueberry harvester. We wonder what this will do to farm jobs and landless farm workers.

Obviously this is a thriving fazenda. Equipment is good, technology appears modern, and the workers seem to be living well. But we have learned little. Our visit has been so brief and our unasked questions so many. Who is the farmer here? The owner? He lives in Tokyo. The manager? He hasn't identified himself. The workers? They all seem to have disappeared. Such explanations as we receive come from fellow visitors, English-speaking Brazilian economists. On reflection we conclude that anonymity is a characteristic of a fazenda and perhaps we expected too much. Besides, we are a group; farmers respond better to an individual than to a busload of people. In addition, there is the language barrier. These problems stay with us throughout the trip.

                          *   *   *

We next go more than one thousand miles north to the Amazon basin, the largest well-watered but undeveloped area of the world. The Brazilian part of the Amazon, measured at its widest, stretches about fifteen hundred miles from north to south and an equal distance from the Peruvian border to the Atlantic Ocean. The Amazon River itself is so enormous that it can hardly be comprehended. It carries more water than the Mississippi, the Nile, and the Yangzi together.

For a long time Brazil developed its agriculture only along the Atlantic coast, as did the American colonies in the early years. Brazil has mountainous country between the coastal area and the interior, as does the United States. But as the Brazilian population increased, the need for food rose, and the desire for economic development grew. There came strong interest in extending agriculture into the new area. There are three general regions with development potential: the Mato Grosso in the southwest, the savanna area of the center, and the Amazon basin of the north.

There is an enormous controversy whether the Amazon basin can be brought into agricultural production. Among those who agree that this is possible, there is another argument whether it is desirable. And those who agree that development is possible and desirable are divided on how it should be done. The problems are vast. The soil is largely unmapped and varied in its potential. The land is quite fragile once the forest cover is removed. There are the problems associated with remoteness, poor transportation, hostile Indians, and tropical disease.

We see half a dozen sharply contrasted efforts to develop agriculture here in the Amazon. All of these are in an early stage.

We begin at Belém, near the mouth of this great river. We go out to the country around Castanhal, a town on the Anhangapi Highway, some distance from Belém. We turn off the highway onto a rough narrow dirt road that is hardly more than a slash through the jungle. Suddenly we come to a clearing where we find a dairy farm, recently wrested from the forest. The name of the farm is Piti Mandena. The owner and operator, whom we meet, is Olivier Silva da Magales, on his horse and in his working clothes. We also meet two of his hired men, likewise mounted. The forest was cleared by hand a few years ago. Jagged stumps and some surviving trees dot the landscape. Grass has started, but it is rather uneven. The farm is in production. There are about sixty or seventy-five cows, of mixed native and Brahmin blood.

Mr. Magales borrowed his money from the Banco da Amazonia, a government-supervised institution that finances agricultural development. Obviously he has launched this enterprise on a shoestring. His home, which we see, is plain but adequate. There is no housing for the cattle, only a milking area.

Cows are milked once a day by hand in the very early morning, before the temperature rises. Milk goes to market in cans, warm, fresh, and unpasteurized, on a pickup truck. The market is the closest town, Castanhal. The practice is to take as much milk from the cows as the market calls for, leaving the rest for the calves. This is a method of supply management once practiced on the American frontier.

Grass and dairy cattle are usually thought of as best suited to temperate latitudes. Here is a dairy operation almost exactly on the equator. We Americans have some misgivings. But Mr. Magales is resourceful and determined; he thinks he can make it. This is an authentic family farm, big enough to succeed if things go well.

*   *   *

We next go about three hundred miles west to Altamira on the Trans-Amazon Highway. This road runs east and west, south of the river, giving access by land to the remote interior. The road is remarkably straight, slicing through the jungle with cuts and fills. The terrain is definitely rolling, with upland soil, not the level alluvial plain some of us had expected. The road surface is dirt, with sometimes a little gravel, badly washboarded. The red dirt rises in dust that gathers on head, hands, and clothes.

Altamira is on the Xingu River, a large and fairly clear stream that enters the Amazon from the south. The site was chosen because the soil was judged to be productive. Accessibility was of course a main factor; the Trans-Amazon Highway was intended to open the area and has done so. There are twenty-five hundred settlers in the Altamira area. The town is booming as were the American towns during the westward movement. Buildings are going up, machinery clatters everywhere, pipes and lumber are stacked in great piles, and workmen hurry about. Clouds of dust hover over all.

Settlers have been brought in from the impoverished northeast, where famine received worldwide attention some years ago. Some settlers are from the south-central area. Others are from the *favellas*, the slums surrounding the coastal cities. All are volunteers. Many have had no farm experience. The settlement is only two years old.

The settlement is under the supervision of the National Institute of Agrarian Reform (INCRA). We are told that the development plan is original with the Brazilians, but it bears a strong resemblance to the Israeli system. Settlers are housed in clusters of simple wooden homes, ranged in semicircles, about fifty to a group, to facilitate the providing of social services. Gardens adjoin the homes, with pineapple, manioc, coconut, banana, and papaya. Fields, which still look rough, are within walking distance. They have been partly cleared and planted. The government helped fell some of the larger trees, but much of the work was done by the settlers, with ax and saw. Farming is to be labor-intensive, not heavily mechanized. Only part of the cleared land is in crops at this stage, and the crops—rice, sugar cane, and corn—are uneven in their progress. We are told that the land is potentially productive and that it responds well to fertilizer.

There are various clusters of these settlements. Half a dozen or so make a complex that is served by a larger unit. At this central location are health services, schools, repair shops, stores, warehouses, marketing facilities, and government offices.

Settlers buy the land on long-term contracts at low prices with subsidized interest. Half the land must be left in forest. Farmers work as individuals but cooperate in buying and selling. The government must carry them at least for a time, until they can become self-supporting.

A sugar mill is going up; when it is operative, the settlers will have a market for cane. The government plans to provide an outlet for storable crops like rice and beans. Markets are too distant and transportation too poor for perishables.

About eighty local Indians had to be removed from the area and resettled to make room for this development.

"Was this hard to do?" we ask.

"It's not completely done," is the evasive answer.

This settlement is a mixture of private enterprise and mandatory cooperation, with government underwriting. It is therefore quite different from the fazenda at Campinas and the family farm at Castanhal. Our dialogue is only with the government people, so we don't learn how the settlers feel. It is obvious that the Brazilian planners are not placing all their hopes on a single farming system.

* * *

We go on to Manaus, nine hundred miles up the Amazon, at the point where the Rio Negro enters. We hear of but do not see a location near Itacoatiara where there is a settlement of Japanese farmers. The report is that they have small intensive farms, producing vegetables, fruit, poultry, and eggs for Manaus. Black pepper is one of the crops. They do well, we are told. This is not hard to believe. Japanese farmers have a reputation for being skilled, industrious, and frugal.

*   *   *

A little farther on we visit a cooperative farming project begun five years earlier, primarily for the Indian people in the area. It covers fifteen square miles and originally had 225 families. The Indians live, as they have for centuries, along the streams throughout the region. If a man is accepted by the cooperative, he is given a tract of land. Pineapple, manioc, and rice, as well as fruits and vegetables, are the principal crops. Produce in excess of family need is assembled at project headquarters and taken to Manaus for sale. There is a sawmill, a rice mill, and a manioc-processing plant. The effort is to change the pattern from roving to sedentary. The leader went to Israel to study the kibbutz and moshav systems as a basis for the project.

Medical service is provided. The health problems in order of importance are malaria, leprosy, and nutritional deficiencies. A school serves 250 pupils. A Christian mission, international and ecumenical, has been established. We meet the Reverend Mr. Butler and Mrs. Butler, wonderful people. A small chapel serves the community, most of the support money coming from the American contributions.

ABCAR, the Brazilian agricultural extension service, is in charge. A strong effort is being made; the project has seventeen full-time workers. We see the headquarters and the government personnel, but not the members of the project.

Despite all this effort the project is failing. The original 225 families have shrunk to 80. The project almost folded last year. Those of us who are from the United States know the difficulties of trying to transform native Americans from wanderers to settlers, from hunters and gatherers to cultivators. Many centuries were required for our European ancestors to make the transition. It seems doubtful to us that any farming system could accomplish the desired changes in five years' time.

*   *   *

Not far from Manaus is the Forest Research Experiment Station, which we see. The Amazon basin is the world's largest forest; trees grow wondrously well. From an economic standpoint, however, there is a problem. On a given acre of native forest there are many different species, some of which are quite valuable, though there may be few such trees in a given tract. The majority of the species are without commercial value. To get at the valuable trees almost the whole forest must be felled.

So the idea is to propagate the desired species in a monoculture planting, to grow trees like a crop. The experiment station is working on the selection and breeding of species, propagation of nursery stock, planting techniques, and fertility practices. Some new knowledge has been developed.

Private investors have acquired extensive holdings here; the land is available at low cost. Some of these holdings are being prepared for monoculture planting. The first step is to clear-cut the entire area, then plant the nursery stock. We drive past mile after mile of felled forest, the trees lying crisscross as if they had been leveled by a giant whirlwind. This is on hilly terrain, subject to erosion. It looks, to me at least, like an ecological disaster. Bereft of its forest canopy, the soil is pounded by rain. Exposed to the hot sun, the humus quickly oxidizes and disappears. Having lost its humus and its absorptive power, the raw land rejects the rainfall. Water pours down the slope, carrying the soil with it. Gullies appear, and silt accumulates below. The soil quickly hardens, losing both its structure and its fertility.

The newly planted trees are not yet sufficiently grown to blot out the damage. We are told that, given a little time, the trees will grow and form a canopy. The soil will be protected as before. Maybe so.

Bringing tropical forest land into agricultural production doesn't need to be this destructive. I have seen it done in Malaysia with good conservation practices.

We are told of, but do not see, an enormous project on the Jari River, which flows into the Amazon from the north, several hundred miles from the Atlantic. This project is the enterprise of Daniel K. Ludwig, the American shipping magnate. Mr. Ludwig foresees a growing need for paper. With the drawdown of the world's forest area, he anticipates a strong market for pulp. Trees grow rapidly on the equator, with eighty inches of rainfall a year. Ludwig is felling the native forest and planting trees from which to make pulp. Ludwig's silviculture methods, we are told, are similar to those we see at the

produce food for home consumption or for sale. The income of a family depended on the amount and kind of labor supplied by that particular family, not on the ownership of assets. Private property had been eliminated. Family farming, the tradition in China for some five thousand years, had been abolished in less than a decade, and socialism had been established. The peasantry embraced the change with astonishing speed and enthusiasm. In Russia the revolution was urban-based and later was pushed into the country against the opposition of many of the peasants. In China, the revolution had a firm and early agrarian base.

In 1958 Chairman Mao launched his Great Leap Forward, a push toward utopian socialism. Huge communes were formed, which were to plan the operations of five thousand or more families. Income distribution approached equality, incentives practically disappeared, private plots were abolished, eating was in community mess halls, and the communes were forced to take on nonfarm enterprises such as small backyard steel furnaces. The effort was a failure: food production fell, famine struck, and most of the communal features of the Great Leap were soon rescinded.

From about 1961 to 1966 agriculture and the nonfarm sector made considerable progress.

But the agrarian earthquake had yet another aftershock: The Great Proletarian Cultural Revolution, from 1966 to 1968. The purpose of this undertaking was to rekindle revolutionary enthusiasm and "destroy outdated counterrevolutionary symbols." Groups of young activists, called Red Guards, disrupted town and country. The resulting chaos wiped out much that had been gained during the previous five years. The Cultural Revolution was conceded a failure and was countermanded.

These things we learn mostly from books before our journey to China. A traveler should spend about as much time in study, before he takes his journey, as he does in travel itself.

The System

Beijing, September 14, 1982. Here we are in China, having received the necessary approval and clearance. We spend four or five days visiting high agricultural officials, who respond readily to our questions. Following is a brief summary of China's agricultural system;

one must understand the system if on-farm observations are to be meaningful.

The present institutional form for Chinese agriculture was laid down in the Third Plenary Session of the Eleventh National Party Congress in 1978. This action, the full significance of which the Western world is just now beginning to perceive, provided production incentives for Chinese farmers that in some respects resemble those available to American farmers. Herewith is a brief summation of the consequences of the 1978 action.

1. Prices. Official prices had been far below the free-market level. These official prices were increased so that the gap between the official and the free-market price was reduced. After being raised, official prices still average 30 percent below free-market prices. When the official farm price of food was increased, retail food prices were held to earlier levels, requiring a large and growing subsidy.
2. Peasant initiatives. The farmers themselves were given considerable decision-making power as to what crops would be grown.
3. Responsibility system. Individual households could enter into contracts to produce agreed amounts of various crops, based on history. For production in excess of the quota, the price would be 50 percent greater than the official price.
4. Private plots. Farm families were promised that each could have the use of a small plot of land, the production to be used at home or for sale in unregulated markets.
5. Sideline activities. Farmers were assured that they could engage in activities outside the collective, such as handicrafts.

With this reform, the structure of agriculture became thus layered:

National level. Communist Party leadership develops policy.

Commune level. The communes disseminate party policy throughout the country and have responsibility for administrative affairs. There are something in excess of fifty thousand communes, having an average of approximately three thousand families. Land is owned by these collectives.

Brigades. The brigade is essentially middle management, responsive to policy directives from above and farmer interests from below. The average commune has about fifteen brigades. Each brigade is typically centered in one or more villages.

Production teams. A production team, which might average 150 people or thirty families, is the basic accounting unit at the grass-roots level. The team contracts with individual families for production of specific crops and livestock, undertakes improvement projects, and keeps records. It elects its own officers.

Families. The "responsibility system" has given families semientrepreneurial status. They look to the production team as their means of relating to the commune.

Individuals. Costs and returns are generally socialized within the family. But the individual does have a measure of existence separate from the family, the team, the brigade, and the commune. He may be given greater or less individual status by his family, and he may be the subject of special action by the authorities.

Chinese communes are cooperatives in the sense that if returns are above outlay, the net proceeds are received by the members.

There are a few state farms, run by the government, the workers being paid straight wages, like factory workers. State farms occupy about 4 percent of China's agricultural land.

A unique feature of the Chinese system is that most of the agricultural officials are of peasant background and so reflect farmer interests. This is in contrast with the officials of many other Third World countries, who are of urban origin and turned to agriculture after having been disappointed in an effort to study law, medicine, science, or business.

At the very top Communist Party membership is total. Party membership diminishes rapidly at progressively lower levels. At Zhouxi People's Commune near Shanghai, which we visited, 3 percent of the people were party members.

The Chinese agrarian revolution was a mighty orgasm. Conception occurred. The child was delivered, a live birth.

We ask a high Chinese official whether the agricultural system that has emerged is essentially stable. He pauses. Then he gives this enigmatic reply: "There may be some further changes."

Self-Reliance, Hard Struggle

Yuan Jia Brigade, Shaanxi Province, September 15, 1982. North China is mostly wheat, corn, sorghum, and millet. South China is mainly rice, vegetables, fruit, and sugar cane. East China is a mixture of industry and agriculture. We hope to see farms in all three of these areas. We start in the north.

We go to Shaanxi Province in north central China, about five hundred miles inland from the China Sea. It has about the same latitude and rainfall as Kansas. The capital of the province is Xian, with two and a half million people. It is an ancient city having been for eleven centuries the capital of all China.

The soil here is technically known as loess. It was brought in from the north and west by winds that swept out of the Gobi Desert many thousands of years ago. In some places this loess soil is three hundred feet deep. It is a light brownish yellow, a mixture of fine sand and clay, easy to till and naturally fertile.

The soil is erosive. Strangely enough, this is less of a problem for the erosive area than it is for the country downstream. The loess is capable of producing good crops throughout its entire profile; if topsoil is eroded, the lower stratum can be made productive. But the eroded silt gets into the river and drops when the stream slows in the eastern lowlands. The riverbed thus rises above the plain; higher and higher banks must be raised to confine it. When the river breaks out of its banks, as it may at a time of excessive rain, there is great devastation. The Yellow River, chief bearer of silt, has earned the name "China's Sorrow." The problem is complicated by the fact that injury (downstream) and remedy (upstream) are not in the same hands. Erosion control is complicated by the fact that much of the soil loss occurs on the uncropped grazing lands of the interior, where rainfall is limited in total but can come in violent storms. Better tillage cannot help these untilled acres.

Mountains that range generally from east to west divide Shaanxi Province into a series of valleys, the largest being the one in which Yan Xia Commune with its Yuan Jia Brigade are located.

Farming here is very old. We see a neolithic archaeological site named Banpo, east of Xian, discovered in 1953, where there was settled agriculture six thousand years ago. The people lived in earthen houses, built on frames of wood. Hoes, sickles, and seeds of millet were found. The pig and the dog had been domesticated. We learn of a man named Hou Ji, maybe mythical, who allegedly taught agriculture to the people five thousand years ago. Northwest Agricultural College at Wu Gong, forty miles from Xian, is named after this man.

We drive out to Yuan Jia Brigade, which is about thirty miles northwest of Xian. Wheat, which was a record crop, has been harvested. Corn, also a good crop, is beginning to ripen. Cotton is ripening, but the crop is poor as a result of late rains. This is unusual; average rainfall is only twenty inches per year. There is millet, and there are a few fields of ripening rice. Topography is gently undulating. There are mountains in the distance. Land has been leveled and terraced, irrigated with water from the Wei River, tributary to the Yellow. There are some nonirrigated lands. Fields are fairly large, sometimes tilled by teams of people and sometimes tended, as to planting, weeding, and harvesting, by individual families who have identified tracts within the larger field. In the fields are many people, a few horses and bullocks, and almost no tractors. Homes are clustered in villages, surrounded by earthen walls.

The people of the past dug into the earth to bury their emperors and lesser officials, then raised earthen mounds to mark the tombs. These mounds, varying in height from 15 to 150 feet, number some 160 in the vicinity of Yan Xia Commune alone. People farm around these mounds, leaving them undisturbed, despite the treasures they may contain. Such is their respect for the dead—and for the authorities.

The farmers of Shaanxi Province are people of the Good Earth. They live with and use the earth in manifold ways. Not only did they bury their emperors in it; two thousand years ago they used rammed earth as the core to build the Great Wall. The people dug caves into the loess cliffs and made homes; they still do. They form the earth into pots and fire them for home use; they make earth into bricks; they build their houses of it. They level the land into terraces for irrigation. They make earth dams to impound water for the crops. They dig into the earth to build canals to transport water to their fields; they build earth bunds to contain the water; they build earthen embankments to confine the river. They nourish this land with compost and nightsoil. They till the dirt of the fields with spade and hoe. It has not been easy. The land of China is sometimes level and fertile. But much of it is steep and hostile to the plow. There perhaps are no other people who so work with and love the earth as do the Chinese.

As with land, so with water. The Chinese manage water with great skill. They were irrigating land as early as 2200 B.C. In 214 B.C. they joined the Yangzi and the Pearl rivers (and thus Guangzhou [Canton] with the north) through the Li River. Demonstrating engineering skill, they devised a great variety of ways for lifting water. Drainage and irrigation are opposite sides of the same coin; they practice both on the same land. They have a flood control system that, with the possible exception of Egypt's, must rank as the greatest in the world. On their intensive vegetable plots, with shallow-rooted plants, they manage so as to keep the water table at a constant and optimum level, about a foot below the surface. They so lead water from higher to lower levels that it walks, not runs. Some of these things they have been doing for five thousand years. In management of land and water resources, Chinese farmers rank with the best in the world.

We Americans, who have worn out and eroded much of our land in a period of two hundred years, have some things to learn from the Chinese.

We arrive at Yan Xia Commune and stop at the headquarters of Yuan Jia Brigade, one of the twenty-six such units in this huge complex. The buildings, which are of brick and are substantial, lie at the foot of an ancient tomb, a mound thirty feet high, which is neither attended nor despoiled. Yuan Jia brigade has forty-five households, altogether 202 people, with a work force of 90. It has about seventy acres, or about one and a half acres per family. We meet Ma Hui Lian, leader of the brigade and himself a farmer. Mr. Ma is thirty-one years old. He is slender, taller than most Chinese, with a shock of unruly black hair. He has a brisk, energetic manner. He has had twelve years of schooling, is married and the father of one child. He was elected a member of the five-person brigade council four years ago. This year it is his turn to be leader. Average age of council members is thirty years. Mr. Ma tells about the brigade, translation being provided by our guide. The dialogue is punctuated by sips of tea.

The people of the brigade are not native to this area. Their forefathers came here from the east several generations ago. The Yellow River had broken out of its banks and flooded the countryside, destroying the crops and causing famine. Some of the refugees came to this place, which was then barren and unused. It was located in the outwash of a stream that came down from the hills, carrying with it much rock and sand. This was the only unoccupied place the refugees could find in which to make a new start. They made their homes in caves dug into the loess cliffs. They obtained a precarious livelihood from the unproductive land.

After the revolution these relative newcomers were put together into this brigade. But until about 1970 things went poorly. Leadership

was not taken seriously. The brigade owned only one horse-drawn cart
and five cattle.

Then came a new and vigorous brigade leader, a man named Guo
Yulu. Mr. Guo shifted responsibility from the aged people to young
men who had some education. He inspired the brigade members to
make changes.

"What changes?"

"We originally had 106 different scattered plots; we collected these
and made new boundaries. We dug up the stones from the fields and
used them to build the borders. We leveled the land so it could be
irrigated. We dug "wells with big mouth" for irrigation, sometimes
twelve feet wide at the top, tapering toward the bottom and as much as
seventy-five feet deep, dangerous work. We got electric pumps to lift
the water. We made compost of night soil, animal manure, and
cornstalks. We gathered sheep manure in the mountains and worked
all this into the soil."

"You must have thought there was no end to the work."

Mr. Ma then retells Chairman Mao's story about the foolish old
man and the mountain. An old man had a house; a mountain
obstructed the way to and from it. So he began to dig at the mountain
and carry it away. "You foolish old man," his friends said. "You will
never be able to move the mountain."

"I have a son and a grandson," said the old man. "There will be
many generations. In time the mountain will be moved."

We sip tea, contemplating this. Is it a piece of satire? Is it just a
fanciful tale? Does it have a moral? Our inability to grasp the nuance
shows how far the East is from the West.

It impresses us that Mr. Ma attributes the improved conditions of
the members to hard work and to the leadership of Mr. Guo; he makes
no mention of national policies.

"What are your grain yields now that you have improved the
land?"

He gives it in Chinese terms, 825 kilograms per mou, which turns
out to be more than 5 tons to the acre, very good indeed. This is the
per-acre yield from multiple cropping. They grow approximately
three crops in two years: wheat, corn, millet, and cotton phased in no
rigid sequence. Wheat is mostly for their own food use; corn is mostly
for sale, both for feed and human food. Five tons per year on irrigated
land compares with a United States average corn yield of about three

tons per acre based on one crop per year, mostly rainfed. The reported 5 tons of grain per year for Mr. Ma's brigade compares with China's national average of 1.6 tons of rice per acre. Obviously we are seeing an outstanding farm.

"Please explain the field operations."

They have three specialized groups within the brigade, each with a particular area of land in its responsibility. Contracts are negotiated with the individual farmers. Because the brigade is a small one, it is the accounting unit. That is, it deals directly with the farmers, keeps records of production and labor, and transmits compensation to members. If the brigade were larger, there would be regular production teams below the brigade, acting as the accounting units.

If the farmer exceeds his contract, he gets paid a premium. This premium is divided between the individual farmer and the various levels of the system.

Compensation to the families is also based on work points, which are scaled according to the skill and energy required. There are more work points for driving a tractor or digging a well than for sorting cotton. We learn that the best workers earn twice as many work points and therefore receive twice as much income as the poorest workers.

Each family in the brigade has a small plot on which to produce food and possibly keep a pig. There are exchanges among families, but almost no sales from the private plots; they are too small. There is some collective land, available to all, used to produce vegetables. Handicapped and elderly people work on this land; produce is sold to members of the commune.

"What is the role of the women in your brigade?"

"They work along with the men. They are not as strong and so cannot do some of the heavier jobs. They take care of the home, work in the private plots and the fields. They are taking over more of the responsible jobs." We can testify to the truth of this last statement. At one of our stops we meet an able woman, Madame Wang Shun Ying, who is vice-president of her commune. But we notice that the women do indeed perform heavy work and that most of the prestigious jobs, like tractor driving, even though not physically demanding, are performed by the men.

"Do you have sideline activities, besides the farming operation?"

They do. The brigade has both a brick and a lime kiln. There is a team of construction workers. They have a transportation team with five trucks. They keep a small herd of cattle for manure and to produce young animals for breeding purposes.

"What are the staple foods here at this brigade?"

"Wheat is our main food. We eat it as noodles or steamed bread (steamed bread is in doughlike chunks, chewy and eaten hot). We eat some corn, some sorghum and millet. For vegetables, cabbage (this is Chinese cabbage, stalky, with long loose heads), eggplant, onions, potatoes, and radish (huge, long, and white). For fruit, at this time of year, apples, pears, and persimmons. Our meat (scarce by American standards) is pork, fish, and chicken."

We note the almost total absence of beef and dairy products.

After the interview we visit a number of homes. The noon meal is being prepared: a hot soup, chiefly of cabbage, potatoes, and onions. Fresh steamed bread will be served. The beverage is tea.

"Tell us about your living conditions."

They are building new homes, which we later see. They are like two-story row houses, well-built brick homes with inner courtyards. They are similar to the homes we see on other communes, in the east and south. Building is done by the members, using the brigade's own brick and lime.

We learn that the brigade plans to build a retirement home, "home of respect for the aging."

There is no place for religion in this system. A few of the older people may retain some vestige of Buddhism. This is not opposed. The expectation of officialdom is that this remnant of faith will die when the people do.

"Do the young people want to leave the brigade and go to the city?"

Not since things improved, says Mr. Ma. The workers in this brigade now earn one hundred yuan (roughly fifty dollars) per month, an improvement over earlier times. Formerly the young people left. In 1980 three farm-leavers returned.

We have a look about. There is a four-wheel, thirty-five-horse-power tractor, but we see no tractor-drawn equipment. Perhaps the tractor is used for pulling wagons. We see tractors on the road, but few in the field. The cattle are in good flesh, tethered, and fed on forage brought in by handcart.

We see only some elderly people, some children, and a few women cooking mid-day meal. The workers are in the field or at some non-farm job.

This brigade is no doubt better than most. We do not criticize the Chinese for letting us see their best; we do the same when foreign visitors come. National figures more or less agreed to by unbiased statisticians confirm the increase in production in recent years and the reduction in disparity of income, though the improvement is less dramatic than at the Yuan Jia brigade. We travel by car, rail, and boat outside the fertile valley farms prescribed for our visits and so get some feel for the average as well as the best. Seeing the farms the officials wish to show gives a reading of what the Chinese think of as desirable; seeing the countryside in an unprogrammed manner provides a rough measure of the gap between the desired and the actual. The gap is considerable.

The Chinese are a no-nonsense people. They are purposeful, energetic, and disciplined.

There is a huge sign at the head of the street in the Yuan Jia Brigade:

Self Reliance
Hard Struggle

The sign sums up the story of this brigade. Indeed, it summarizes the story of the Chinese people.

Balancing Agriculture with Industry

Malu Commune, September 21, 1982. In virtually every country the farm population is a diminishing share of the total. This is largely the result of industrialization; with modernization of agriculture and industry the increments of economic growth and job creation are greater for the nonfarm than the farm enterprises. The question is whether these increments of industrial growth and job creation are to be located primarily in the urban or in the rural areas.

In America, from colonial to recent times, most of the nonfarm jobs were created in the population centers. But more recently the United States, relying mostly on individual decision-making and voluntary methods, is shifting its new industrial growth to the rural areas, providing part-time or full-time jobs for members of farm families. The nonfarm incomes of American farm families now total larger than their earnings from agriculture. The 1980 census revealed that even though the number of farm jobs continued to diminish, this was more than offset by the growth of nonfarm jobs in the countryside. For the first time since the original census of 1790, the rates of population growth and job creation in rural regions exceeded those in the urban centers. In the United States the shift of population from rural to urban areas had been reversed.

How are these matters working out in China? China has a centrally planned economy. She intends gradually to modernize agriculture, which means replacing human labor with machines. She intends to industrialize, which means job creation. Will she build most of her new increments of industry in her population centers, as we first did in the United States, and thereby swell the growth of her already enormous cities? Or will she try to balance agriculture with industry, building new jobs in the countryside to absorb, in place, the rural workers released by the transition from hoe to tractor? How does this centrally planned economy deal with this question?

The best place to study this subject is the Shanghai area. Shanghai, on the coast in eastern China, is in the industrial heart of the country. A city of eleven and a half million in the midst of a rich farming area, it reflects the interface of farm and nonfarm. So we go to Malu People's Commune near Shanghai to learn what we can.

Malu Commune is twenty miles north and west of Shanghai. This is the delta of the Yangzi. Soil is the rich alluvial gift of the interior uplands. Rainfall is forty-five inches per year, supplemented as needed with unlimited irrigation water from the Yangzi and its tributaries. Temperature is mild. Plant growth is possible throughout the year. Multiple cropping is standard practice. Per-acre yields are very good. The latitude is similar to that of Mobile, Alabama. The Shanghai area is one of China's garden spots.

The city of Shanghai keeps sprawling out into this garden. Factories and apartment houses spring up amidst the beans and cabbage. Cranes and derricks tower above the rice fields. New buildings creep into the countryside like the scouts of an advancing army.

Malu People's Commune has more than eight thousand families, and a total population of about thirty-two thousand. The labor force is nineteen thousand. There are fifty-five hundred acres of cropland, all of it irrigated. The farm has seventeen brigades and 155 production teams.

We meet Fan Tiah Min, who is a director of the office of the commune. Mr. Fan is responsible for records, accounting, and analysis. He is one of the five officers who collectively guide the operations of this huge place. He holds his post by appointment. He is probably in his upper thirties: tall and slender, with short dark hair. He wears a white short-sleeved shirt and dark trousers, typical and correct warm-weather garb for one in his position. More formal wear would be the blue party uniform; less formal, for the workmen, would be the knit sleeveless shirt, with either trousers or shorts. Mr. Fan is a native of this commune, an agricultural graduate of a middle school, which means he has had twelve years of schooling. He was a teacher of mathematics in the commune before getting into administrative work. He is married and has two sons, aged twelve and nine, now in school. College is planned for them. Mr. Fan is frank and friendly.

The chief agricultural enterprises, he tells us, are rice, cotton, and wheat. Additionally there are rapeseed, mushrooms, sweet potatoes, and vegetables. There are substantial livestock enterprises: pigs, eggs, broilers, ducks, geese, dairy cows, rabbits, and deer (for their antlers, which are ground and used in traditional medicine as an aphrodisiac). They also culture pearls and produce freshwater fish. We later see most but not all of these enterprises.

The farm is modernizing. There are 232 tractors. Most of them are two-wheeled walking tractors, the rugged simple Chinese-made one-cylinder water-cooled twelve-horsepower machines we have seen so often. These tractors are used to till the land quickly after the early crop of rice is harvested, so that a late crop can be planted in time to mature. The tractors replace much hand tillage. China is shifting toward four-wheel thirty-five-horsepower tractors that will expedite the turnaround between crops even more.

Tractors are hitched to wagons, pulling great loads of grain and vegetables to market on the outskirts of Shanghai. Trucks as well as tractors supply transport. Formerly much produce went to market on carrying poles slung over the shoulder, requiring enormous amounts of labor.

We see a pumping station, of which there are forty. They are electric-powered, replacing the old manual methods of lifting water.

The commune's pig production is modern large-scale confinement style, requiring much less labor than the old system of a pig per household. The same is true for eggs and broilers.

Threshing is done with a slatted rotating drum, powered by a small motor. Formerly threshing was done manually by beating the grain against a board.

Herbicides are coming in. They will replace in part the enormous labor involved in hoeing and weeding.

The substitution of mechanical power for man power and the advent of new methods have gone much farther and faster on Malu Commune than on other communes in China. Nevertheless, the trends under way here are discernible, in their early stages, in many parts of the country.

At Malu Commune agricultural production has increased, but labor requirements have diminished. Of the labor force of nineteen thousand, only eleven thousand are needed for the production of crops and livestock.

What of the other eight thousand workers? The government will not allow them to go to Shanghai and offer their labor, except under very special circumstances. So there must be nonfarm employment on the commune. Mr. Fan tells us about these enterprises. They make some of the simpler farm machinery. They make hardware such as wrenches and clamps. They make furniture. Bamboo is made into a great variety of products. There is a candle factory. They dry, pack, and sell garlic and mushrooms. They make lemon extract of the sweet potatoes they grow themselves, a sophisticated form of food technology. They make monosodium glutamate. They make garments and toys. They press oil from soybeans. Altogether, says Mr. Fan, they have forty-eight minifactories and produce one hundred different products. Many of these products are sold under the commune's brand name. Some of them go into export either under the Chinese label or under the label of the importing country.

In addition to this factory production, some of the members engage in cottage industry, producing embroidered work and handicraft articles.

The 40 percent of the commune's workers who labor thus in light industry generate 76 percent of the commune's revenue. Income from the industrial enterprises has made this a wealthy commune.

Another commune we saw in the Shanghai area, Zhouxi, had 45 percent of its labor force in industry. These communes are by no means typical; Shanghai offers opportunities not generally available.

The rate of farm mechanization is much greater in the Shanghai area than it is for the country generally. And the opportunities for industrialization are likewise greater. Nevertheless, at each of the six communes we visited, light industry was a substantial user of labor and an even more substantial contributor to income. We learn from a government official in Beijing that even in the remote border regions nonfarm income makes up 10 percent of the total income of the communes.

We people from the United States tend to think of Chinese communes as agricultural enterprises, because they produce most of China's food. Consequently we feel something of a shock on arriving at the headquarters of a commune and seeing what resembles the county buildings in an Indiana town, with no agriculture in sight. It looks like a small city. It *is* a small city. The communes are geographic entities, and they incorporate the villages within their borders. That means they include the shops, the factories, the schools, the hospital, the law-enforcement function, and the job of tax collection. Thus it is inevitable that a commune have at least some nonfarm activity.

A special feature of the Chinese system is that there is limited movement of people from one commune to another, or from farm to city. In an open society free movement of people helps equalize economic opportunity. The people go where the opportunities are. The Chinese system precludes this form of self-adjustment. Consequently the state must undertake the whole equalization process, pumping capital into the less developed areas and building new opportunities for the disadvantaged. Otherwise the more fortunate areas would greatly increase their income advantage over the less favored. Despite the effort to equalize incomes, regional disparities are already occurring and widening. The reduction of regional income differentials is very difficult with China's obstacles to population movement.

We visit the candle factory at Malu Commune. A Western industrialist would criticize it for inefficient handling of materials, wasteful use of space, and use of hand labor for operations that could be mechanized. A man with a drill press bores holes for wicks, one at a time, all day long. Candles are being individually wrapped by hand. But the enterprise is making a profit. Labor receives about ten or fifteen cents an hour and is glad to get it; American standards of labor use are inappropriate. The workers are dexterous and diligent. They respond well to managerial directions. One reflects on what would happen in the international markets if China should develop her industrial potential and enter the export trade in earnest.

We visit a family that portrays the interface between farm and nonfarm enterprises. This is the family of Mr. Zhang. Mr. Zhang was a worker in a glass factory, an enterprise not owned by the commune. So he is not a member. But he married a commune member, with approval of the governing body, and set up a home in the commune. They have three children. Mr. Zhang, fifty-five years old, is in poor health and has retired. His eldest son, twenty-seven years old, has taken his father's former job, with approval of the authorities, and so has lost his membership in the commune. Mrs. Zhang works six hours a day for the commune, in the garment factory. She often works after hours in the field with the South Malu Production Team, earning work points. The second son, twenty-five years old, works in the field for the same production team, as does the daughter, aged twenty-two. All three were picking cotton when we visited. We later met Mrs. Zhang and the daughter, who came home for noon meal. In addition to these various cash incomes the family produces food for home use in their private plot. The incomes are pooled.

By hard work and good financial management the family has been able to acquire a good two-story house. They own the home, but not the land on which it stands. They have electric lights, running water, radio, and television. They have acquired ownership of an adjoining house of like size into which the older son will bring his bride, a commune member, after the wedding, planned for the spring festival. It seems complicated. But this Chinese family has figured it out. To them the commune together with its industry means the good life.

We ask Mr. Fan how the commune officials decided on whether to launch a new enterprise. Someone makes a proposal, he says, and the top officials consider it. They reflect on such questions as these: Do we have the raw materials? Do we have the labor? Do we have or can we develop the necessary skills? Is there a market? Do we have the money for the necessary capital outlay? If not, can we borrow it? The state loans money at 5 percent interest on promising enterprises. Members of the governing body, mostly of peasant background, weigh the answers to these questions and decide go or no-go much as in a capitalistic society.

The advantages of thus building industrial enterprises into the farming operation are several:

1. It adds to the incomes of the members.
2. It helps prevent migration to the already overswollen cities.
3. It decentralizes industry, which is advantageous both from a developmental and a defense standpoint.
4. It provides flexibility. During peak harvest time, when extra farm labor is needed, the factory can be closed down and the farm labor force correspondingly increased.

The United States is achieving balanced growth of industry and agriculture by voluntary methods. The Chinese are achieving it by central direction. The methods differ, but the consequences are similar. At an early stage of industrialization, the Chinese are trying to avoid the errors we are having to correct.

Strange Job for a Farmer

Dali Commune, September 23, 1982. Ye Ping Wen, deputy chairman of the management committee of this commune, is a tough party man. He and four others provide the collective management for this huge operation just outside Guangzhou (Canton). It has twenty-one thousand households and numbers eighty-one thousand people. Mr. Ye has responsibility for the production of fifty-four hundred acres of rice. He is accountable for the annual production of more than thirty million dollars worth of industrial goods. Mr. Ye also has responsibility for the nonproduction of babies. This last task may seem a strange undertaking for a farmer, but Mr. Ye does not appear to consider it so.

China has a family planning program. We learn about it from government officials in Beijing. We want to find out how well the system functions out in the country. So we ask about it at each of the six communes we visit. The experience reported varies somewhat, but is fairly similar throughout. At Dali Commune we bear down on the subject.

Mr. Ye is a small man. He is brisk, alert, and thoroughly competent. It is clear that his commune is one of the best. We learn that he was from a poor peasant family in this area and has had only five years of schooling. He was quite young when the revolution came, but became an active participant in 1953, as soon as he was able. Party leaders were quick to detect his zeal and his potential. He went to a party school for several months and then briefly to a special school, where he was taught elementary agriculture. Everything else he learned on his own. He began with a production team and quickly became its leader. He has spent almost thirty years in this commune, ten of them in his present post, five consecutive two-year terms. He provides us with technical information about hybrid rice and tells about the industrial operation much as would an American manager of a conglomerate firm. He answers our questions about family planning with the same competence. He has the statistics on population as well in mind as he does those on rice production or industrial income.

"Last year the birth rate at Dali Commune was twelve per thousand," says Mr. Ye. This was better (lower) than the national average, which was nineteen per thousand. But it was not as good as at Malu Commune, which we visited, where the rate was held to eleven per thousand. The death rate at Dali is very low because of improved medical care. Even with this success in reducing the birthrate to about one-third of its historical level, the net population growth was eight per thousand, a little less than 1 percent per year.

"Sixty percent of the births at Dali were first children," Mr. Ye says, citing a standard of measurement that has come into general use. This was close to the national average, but not nearly as good a record as at Yuan Jia Brigade, which we visited. There only two children were born last year among forty-five households, both children first babies.

Mr. Ye, in his early grandfather years, is the father of five children. This large family is no embarrassment to him because the children were born before 1970, when family planning first appeared as a clearly stated party objective. In China, the standard for current deportment is the most recent pronouncement of the party.

The circumstances that led the Chinese leaders to a family planning program were compelling. China has a billion people, four hundred million of them born during the thirty years since the revolution. Improved health care has almost doubled the average lifespan during that period, from thirty-five to sixty-nine years. (The United States figure is seventy-two years.) China is already densely populated; there is only a quarter of an acre of tillable land for each inhabitant.

The United States has about two acres of tillable land per inhabitant, eight times as much as China. If recent rates of population growth are projected into the twenty-first century, the line runs off the chart. Chairman Mao said: "We cannot allow population to grow in a blind way. It would interfere with the economy." All of this is clearly known to Mr. Ye and to others in leadership. It appears to be known to a fair share of the populace.

In 1978 the National People's Congress took action on family planning. They wrote into the constitution that "the state advocates and encourages family planning." Later actions firmed up details. A top figure of 1.2 billion people was specified, cresting about the year 2000, then diminishing and stabilizing at about 700 or 800 million.

There are several initiatives toward the achievement of this objective, one of which has to do with age at marriage. The government recently announced the marriageable age for males to be twenty-two years, twenty years for females. This was a reduction from an earlier figure—reason for the reduction is not stated to us—and resulted in a spurt of marriages, with resulting increase in births. "It makes a problem for us," says Mr. Ye, wearily. Another problem is that during the Cultural Revolution, when restraints were relaxed, there was a surge of births. The millions of women of this cohort are now coming into childbearing age. In 1980, 32 percent of the women were less than fifteen years old.

As to the number of children, the government says that the first one "can come at any time," the second is "controlled," and the third is "prohibited." If unapproved by a responsible official, a child is a "violation."

These edicts are supported by an educational program. The state has control of press, radio, television, schools, and movies. The family planning program is constantly before the people, in strong advocacy terms. Nothing contrary to it is said.

"We have a strong educational program in the commune," says Mr. Ye. Leadership is given at all levels. There are films and lectures. Contraceptive devices of all kinds are explained and are provided free of charge. Men as well as women are brought into the educational effort. The paramedics, "barefoot doctors," work with the people on these programs. Sex between human beings is treated with frankness similar to matters of sex between farm animals. Each couple is urged to sign a certificate, pledging to be a one-child family. Inducements are held out to persuade couples to sign such certificates. Mr. Ye outlines these, which in total are impressive. Cooperating (certified) families receive preference in housing. Nursery care and kindergarten are available free of charge. Extra food rations are available. Private plots are larger for one-child families, and job preferences are given. In some communes, there are cash allowances for one-child families. A mother who signs a certificate to have only one child is given maternity leave up to six months. Meanwhile, she draws the normal work points and so earns the normal income. In a fashion exactly opposite to that of the United States, if additional children are born, the benefits are withdrawn.

Admittedly, some of these provisions are not fully implemented throughout the country.

Peer pressure is applied. Small groups monitor individual menstrual cycles; card files are kept on these matters. A woman who has an unauthorized pregnancy is looked on with reproach in these small groups. This form of coercion comes under the heading of counseling.

Some couples are reluctant to limit the family to one child. There is the desire to have more than one, so as to have support in old age. To overcome this reservation there are "guarantees" for the childless elderly. Mr. Ye outlines them. They are five:

> Guaranteed food
> Guaranteed housing
> Guaranteed medical care
> Guaranteed cash income
> Guaranteed burial expenses

"Are there abortions?" we ask.

"Yes," is the answer. "It is like medicine. Prevention is best, but if it fails, treatment is necessary."

Abortions are not only condoned; they are advocated and performed free of charge. We do not learn here at Dali Commune how many abortions there are. But at one of the other communes we get an estimate, reluctantly supplied, of one for every five live births. Other information available to us indicates that this is a gross underestimate.

In China matters of sex are not flaunted as in the United States. There are no erotic magazines or movies. Dress is plain. The nonsexual aspect of advertising would please the Moral Majority. Concern for the country and the commune are put ahead of erotic interests. Activities for the young are group-centered; young people are discouraged from pairing off. There are almost no automobiles with which to get away from others. Courtship in the Western sense seems not to exist. By all reports teen-age pregnancies are few. An American team of family planning experts reported, after a visit to China, that premarital intercourse was something of a rarity.

We made brief acquaintance with a young Chinese man who spoke some English, and, like many Chinese, was anxious to try it out on us. He said he liked to read. "What do you read?" "Romance," he said. "What, for example?" He thought a while, then said, "There is a story about a girl from Taiwan. She met a young man from the People's Republic. She escaped from Taiwan and came to the People's Republic where she joined him on his commune." We made no further inquiry into this titillating tale.

"Do you have any problems in carrying out the family planning program here at Dali Commune?"

Mr. Ye is frank. "Yes. Some of the older people still believe in large families. Young couples are pressured by their elders to provide grandchildren. There is still a preference for boys over girls. Some people still think the family name and the inheritance should descend through the male line and so want boy babies. Some say boys are stronger and can earn more work points in the field. So if the first child is a girl, there are applications for a second child, hoping it will be a boy."

China has had a pronatal, promale tradition perhaps as long and as strong as any in the world. The Chinese love their children. It is not surprising that difficulties are encountered.

We note by examining national statistics that the family planning program is going better on the communes we visit than is true for the rural area as a whole. All the evidence is that the family planning program goes better in the cities than in the rural areas. Education is higher in the cities, housing is tighter, and tradition less strong. Apart from government policy, children would be an economic burden in the city and an asset in the country.

China has numerous small ethnic groups. The family planning program is not applied to these, perhaps in order to avoid the charge of genocide.

The Population Crisis Committee of Washington reported in September of 1981 the rate of natural population increase in China to be 1.1 percent. This was down from an estimated 3.0 percent during the Cultural Revolution. Had the 3.0 percent rate of growth continued, the population would have doubled in twenty-three years.

We traveled through a number of countries, interviewing farmers, before coming to China. We asked each farmer about family planning. Sometimes the question was not understood. Sometimes it was understood but heard with apathy, or greeted with laughter. Not so in China. Every one of our respondents knew about family planning, took it seriously, and gave it at least oral support.

Mr. Ye tries to overfulfill his rice quota and underfulfill his quota of babies. This dual role may seem odd to a Westerner. But to a central planner it makes much sense. The task is to feed the people, to avoid famine. One way to do this is to grow more crops. Another way is to produce fewer people. The party has chosen to do both. To Mr. Ye, a loyal party member, the decision is accepted without question and implemented in even-handed fashion.

The People's Republic of China abolished the family farm within a decade. In a like period of time it reduced organized religion to a tattered remnant of its former self. Now it has undertaken to socialize the reproductive function, which long has been the most personal and private aspect of human life.

Whatever may be thought of the system of the People's Republic as regards ends pursued and means employed—and I have grave misgivings about both—one must stand in awe of its raw social, economic, and political power.

El Salvador

Subsistence Farmer

Finca Singuil, Ahuachapán, October, 1981. Juan Flores Salinas is fifty-eight years old. His skin, naturally dark, is tanned by the tropical sun to the color of saddle leather. He is lean, relaxed, and remarkably erect for a man who has spent much of his life bent over, grubbing the ground of the volcanic uplands in this sliver of a country on the Pacific coast of Central America.

Salinas intercrops corn and beans as did his father and many fathers before that. He and his son together have two manzanas of land. A *manzana* is about seven-tenths of a hectare, or about one and a half acres. The average farmer of the United States operates 140 times as many acres as does Salinas.

Salinas was a *colono* until a year ago. A colono stands somewhere between a serf and a freeholder. He is not strictly bound to a feudal lord, nor is he really free to launch an enterprise of his own. He works for the owner of the estate on which he lives. Usually there are scores or even hundreds of colonos on an estate, otherwise known as an hacienda and sometimes as a *finca*. This system has been traditional in Latin American for more than four centuries. In exchange for his labor the worker receives a home and a small plot of ground on which to produce food for his family. The owner changes the size and location of the plot according to the changing number and need of the worker's family. The plot may be large or small, good or bad, in accordance with the colono's deportment. When coffee is to be harvested or cane cut, the owner's need for labor obviously comes first. The worker owns no land, little livestock, and no implements except the heavy hoe which is the tool of subsistence for farmers the world around. He barely owns himself. The arrangement is almost a moneyless institution.

That was Salinas's life until a year ago. Now he is a landowner. He pulls a piece of paper from his pocket to show. He is acquiring ownership of this tiny farm with the help of the Salvadoran government, which in 1980 instituted a land reform. El Salvador sought to bring its land tenure system into the twentieth century before entering the twenty-first. Salinas's paper, soiled with sweat, gives him provisional title, which will be final when his land is appraised, the price determined, and his annual amortization payment is fixed. The terms are favorable. The paper is signed by the head of the land reform program and bears Salinas's thumbprint and his mark beside his name.

Salinas has his machete tied to his waist by a thong adorned with leather tassels. The handle is close to his right hand. It is a wicked-looking knife, about a foot and a half long, curved inward at the end and bent so as to strike level when he leans over to cut weeds. It was made from a car fender by a local craftsman and is razor sharp. He uses it not only to cut weeds but also to chop coconuts and, if necessary, to defend himself. He wears it both as a tool and as an article of dress, as a gentleman of former times would have worn a sword.

This man lives with his elderly father, the three of his seven children who are still at home, and his common-law wife. Marriage is costly, and besides, a man might tire of living continuously with the same woman. Salinas has not so tired; he has shared his life with his companion for forty years. Salinas is a *mestizo*, a *ladino:* his ancestry traces both from the Indians who first lived in El Salvador and from the Spaniards who conquered it. These twoblood lines have been blended now for many generations to produce a people remarkably handsome and homogeneous. One of Salinas's sons works with him. This is a three-generation family.

The home in which the Salinas family lives is fairly substantial, better than most. The former owner of the finca, a woman, had more than ordinary concern for her people. The farm she owned, named El Singuil, was eleven hundred acres, fairly large, though there were many much bigger. That was before the farm was expropriated. Salinas learned from his owner that there had been an agrarian reform and that he was eligible to apply for title to land; at her suggestion he did so.

The ecology of Salinas's farming system is good, however baffling it might seem on first observation. To begin with, his crops are well suited to his family's nutritional needs. The corn gives carbohydrates, the beans supply protein, and his cow grazing on a distant tract of untillable land provides him with milk. Sugar cane supplies sweetening. Some fruit can be had without great difficulty in this tropical country. Corn provides more nourishment per acre than any other plant that grows and is the basic food crop in this heavily populated land.

Intercropping corn and beans is an intricate and interdependent system. In April, before the rains, last year's cornstalks are chopped up, by hand, with the machete, leaving the field ready for planting. Planting is in May when the seasonal rains start. Salinas has no tractor and no draft animals; planting is therefore all hand work. His field is too steep for a tractor, sloping at an unbelievable thirty degrees. Some fields are even more steep, as much as forty-five degrees. Land is scarce in this heavily populated country. The basic slope is that taken by dust descending from volcanic eruption. A tractor would literally fall out of the field. Even oxen would have a hard time. Anything other than hand tillage is impossible; the field is strewn with volcanic rocks which range from football size to as big as a house.

Salinas practices minimum tillage, as his fathers always have, partly through wisdom and partly of necessity. If he stirred the soil deep, his land would wash down the slope with the first heavy rain. He prepares his land as he plants it, working up a small area with his grub hoe for each seeding spot.

Now comes the seeding. He has made three major innovations in his agricultural methods, the first of his family line to do so. The first of these he now invokes. He plants hybrid corn, H-5, he tells us in his soft Spanish. This is an open-pollinated variety, selected and fixed from a series of crosses made at El Bataan near Mexico City, the International Center For The Improvement of Corn and Wheat made famous by Norman Borlaug. This is a white corn, with glossy vitreous kernels, having something of a golden sheen beneath the silver, much pleasing in appearance and good for making tortillas. The hybrid yields three times as much as the native, or *criollo*, corn.

The seeding process is fairly standard. The seed is carried in a hollowed-out gourd. Two or three seeds are dropped in a planting. Plantings are separated from one another in the row by the distance between the thumb and the little finger of a man's outstretched hand.

Then comes the second of Salinas's innovations. He puts a small amount of fertilizer beside each of his seed placings. He uses 20-20-0; that is, 20 percent nitrogen and 20 percent phosphate. Potash is not needed. A small matchbox full per seed placing is the standard amount.

When the corn comes up, each hill is thinned to a single plant. Then the hand cultivation and hand-weeding begin.

When the corn is nicely up but before it has made a canopy, Salinas plants beans between the corn rows. They come up promptly, and the two crops grow together.

Now comes the third of Salinas's innovations: pesticides. Beans are subject to all manner of diseases and insect pests. With a backpack duster Salinas applies the chemical. The beans thrive and grow in that combination of rich soil, bright sunshine, and gentle rain that makes El Salvador so productive a country.

When the corn approaches maturity and the ears have drawn from the stalks all the nutrients they will get, Salinas "folds" the corn. That is, he bends each stalk over, by hand, the location of the fold being just below the ear. This has two purposes. It lets the ear hang down so that moisture will not gather at its base and cause spoilage. And it lets in the sun, so the beans will grow better. The corn has been bred to have a thick husk, extending well over the tip of the ear. Thus the damage from birds and the rodents is suppressed. Safe from spoilage and safe from pests, the corn is stored on the stalk until needed. The beans now climb up on the dry brown still-standing corn stalks, and the field takes on a green color.

Then there is the time of waiting for the harvest, during the late summer and early fall. This is a pleasant time, full of anticipation. It is during this time, in October, that we visit Salinas.

Salinas harvests about two-thirds as much corn and beans from his tiny farm as the corn and beans a United States farmer harvests from a similar acreage. And this is on land that a United States farmer would not think of cultivating.

When the harvest has been completed, the cornstalks are left in the field, providing cover for the land. Thus the land has protection from the elements during all but the brief period between planting and the achievement of substantial growth. The intercropping of corn and beans goes on year after year on the same ground. The corn provides the trellis for the beans. The beans, a legume, add nitrogen to the soil to help the corn. The competition between them for sunlight and rainfall is limited, their growth cycles being differently timed. To-

gether the corn and beans provide Salinas and his family with a diet
that is nutritious, low in cost, capable of being stored, easy to prepare,
and delicious to the taste. The essentials of this system were already in
place when the Spaniard Pedro de Alvarado arrived from Mexico into
what is now El Salvador in 1524.

Salinas's farm is near the top of a ridge, reached by a path that
winds up from the valley floor. He is not far from the Guatemalan
border. On the horizon, in three directions, are silhouetted the extinct
volcanoes that are the distinguishing mark of this beautiful country.
Behind and to the right the mountain rises still higher. "How is it that
there is so little erosion?" we ask, through Peter Cody, who translates.
Salinas gives an answer that would be a credit to any farmer. "I take
care of my land," he says, "and with this soil, the rain goes in instead
of running off."

"Do you sell your crop, or do you need it all for your family?"

"I sell some when I need money."

"What do you need money for?"

"Seed, fertilizer, chemical," is the answer. Then he adds, with the
sigh characteristic of the consumer society the world over, "Electric
bill. It goes up." He feels the attractions and penalties of moving from
a subsistence to an exchange economy.

"How much land does it take to support a family?"

"What I have," he says. The fact is inescapable. He is supporting
his family. He earns some money by working off the farm, picking
coffee on one of the neighboring fincas.

"Have things changed for you?"

"Not much change," he says.

"Are you contented?"

He looks first down the slope at his corn and beans, then across the
valley toward the distant greenclad volcanoes. "Yes,"he says, "I am
contented."

Not everyone is contented in this wounded country. On a big farm
nearby we meet a troubled man. He is a self-taught artist, a kind of
amateur Diego Rivera. He shows me, privately, some sketches he has
made of the previous owner of this farm, a man now dispossessed and
living in Miami. The drawings show the former owner as a monster,
squeezing the last drop of blood from his workers. We hear elsewhere
that this former owner helps finance the death squads that harrass the

officials responsible for implementing the land reform. This is a complex country. Almost everything said of it is true for some specific spot, time, or person.

Juan Flores Salinas is a famine fighter who knows for whom he fights—his own family. He makes changes slowly and only after he thinks they will succeed. Survival is the test. His traditional methods must be right; he and his forefathers survived, using them. His margin for experimentation is very small. If he should try something new and it should fail, his family would perish. If an Iowa farmer should try something new and it should fail, he might merely worsen his cash flow position.

But even so, Salinas is making changes. He is leaving his dependent feudal status and becoming an entrepreneur. As he becomes a landowner, his social status rises. He is entering the exchange economy. He has adopted the three most productive techniques of modern agriculture: seed, fertilizer, pesticides. Things are changing, perhaps more than he realizes. But for him, food until the next crop has always been the main problem, and the distant future carries an enormous discount.

India

Farming under Democratic Socialism

Jharoda Kalan, Haryana State, August 17, 1982. This is the time of the summer monsoon. The countryside is green and beautiful even though the rains have been erratic and short of normal. The main crop we see is sorghum in various stages of growth; it is used as forage. The wheat has been harvested, a good crop. This is not rice country. We see some fruit: guava, citrus, and bananas. There are vegetables: tomato, eggplant, cabbage, and okra. We see some kenaf, a fiber crop similar to jute. There is open ground, awaiting the time of planting.

People are everywhere, on the road, in the fields, in the shops. They carry loads, drive burros, and walk about on errands we cannot imagine. They wear the flowing white robes that seem to be out of the Old Testament. Children constitute a large share of the population, the elder tending the younger. The people part in both directions to make way for our car.

Big black buffalo, kept for milk, are all about. They are in good flesh; last year's rains were abundant, and feed has been plentiful. Bullocks are the draft animals. They are white, large-boned, and placid, pulling carts piled high with sacks of grain. Burros are the beasts of burden, carrying enormous loads of cut sorghum to be used for feed. Camels, haughty and disdainful, are used both for draft and burden. They tower high above the wagons they pull.

This is Haryana State, in north India, about the same latitude as Houston, Texas. It is the heart of the Indo-Gangetic Plain, which extends for 1,500 miles in a generally west-east direction. In India the plain is about 175 miles wide, bounded on the north by the Himalayas and on the south by the Deccan Plateau. This is one of the world's most productive areas and is densely populated. It provides India with a large part of its agricultural output. Land is level. We are close to the

Jamuna River, major tributary to the Ganges. The soil is yellow-gray, fertile, deep, loose, and sandy in texture.

We stop at a lounging place in Jharoda Kalan, a village of perhaps five thousand people. this is the slack season, between planting and harvest; farmers gather to visit as they do the world around. We find Pooran Chandra, a forty-five-year-old farmer who is willing to talk. We learn that he owns a small farm, in two tracts, one just outside town and the other some distance off, at the end of a long and very bad road. Mr. Chandra is a Brahman, of the high caste, as shown by the special thread woven into his shirt, which he puts on for the interview. (Maybe he borrowed the shirt; he was naked to the waist when we first saw him.) He wears a dhoti, the flowing white trouser-skirt garment that seems as if it would hobble a man and yet permits freedom of action. He wears sandals and is bareheaded. We learn that he has had education "through the seventh class." A bit of frosty beard is on cheek and chin. His skin is swarthy, and his features are similar to our own. Those who make a study of these things say that his ancestors and ours came from the same place untold thousands of years ago—somewhere in western Asia.

Chandra lives in the village, as do virtually all of India's farmers. The American style, farmers living on the farm itself, is the exception in the world, not the rule.

Chandra speaks Hindi. Translation is provided by B. N. Chattopadhyay, agricultural extension officer, and by V. M. Tandon, of the American agricultural counselor's office.

We go out to see Chandra's farm. The first tract, along the road, is a little less than an acre. Half of it is in sorghum, feed for his animals. The other half awaits the planting of tomatoes and like vegetables, to be sold locally. The distant tract, an acre and a half, is used for wheat and *bajra* (spiked or pearl millet). Thus Chandra's farm is a little more than two acres, which is about the median size for Indian farms. The average American farm is two hundred times as big. Some other striking comparisons: the United States has 2.3 million farms, while India has 82 million, thirty-five times as many. The farm population of the United States constitutes 2.6 percent of the 229 million total; the farm population of India is 70 percent of 684 million.

Chandra grows two crops a year. Vegetables are a cash crop. The wheat is largely for home consumption. What is left is sold. This past year part of the wheat was damaged by rain at harvest time. The

damaged wheat was sold to the government at the official procure-
ment price of 142 rupees per quintal (about $4.20 per bushel). Good
wheat could be sold in the market for about 150 rupees. So the good
wheat went into the market and the poor wheat went to the government. How like the government programs in the United States! The
government acquired 7.7 million tons out of a crop of 37 million tons,
and is now moving this wheat through its Fair Price Shops, of which
there are many thousands in the country. Earlier, we visited one of
these shops. The proprietor, who is licensed by the government, sells
wheat at 1.64 rupees per kilogram (about eight cents a pound) to
holders of ration cards. Each card is intended for a family of five and
authorizes the purchase of eleven pounds of wheat and eleven pounds
of rice a month. Buyers of this wheat, the poorer people, either have
the wheat ground at the village mill or grind it themselves with stone
and roller. In the open market the cost of wheat is 2.10 rupees per
kilogram, 22 percent higher than at the Fair Price Shop, but it is better
wheat.

Chandra grows a new, high-yielding wheat bred by the Indian
Agricultural Research Institute. Using irrigation and a moderate
amount of fertilizer, he gets forty-four bushels to the acre, compared
with India's national average of twenty-one bushels and the United
States average of thirty-three bushels.

On another farm which we had visited, ten miles away, with better
soil, more fertilizer, and better management, we found a man who,
on a demonstration plot and with the help of the extension service, got
a yield of one hundred bushels to the acre.

"Could you step up your yield?" we ask Mr. Chandra.

"I could use more fertilizer. But I don't have the money."

"Could you borrow?"

"If I borrow, I have to give a mortgage on my farm. Then if I have a
bad year, I lose my farm." This cautious man settles for less risk and
less income.

"What is the size of your family?"

He has two boys and five girls. One boy is married. Another works
at a telephone exchange. The five girls are all in school; they work on
the farm when they are not in class. Chandra's elderly father lives with
them, working half-time. Chandra is full-time, and his wife works on
the farm half-time in addition to household duties. Together, the
children do about as much work as a full-time person.

We visit two other farms. One farmer had four children, and the
other six. Since we first visited India in 1958, the country's birth rate
has fallen only slightly, from about forty per thousand to about thirty-
six. The death rate, however, has come down sharply, from about
twenty-five per thousand to about fifteen. In twenty-five years the
population of India has grown from about four hundred million to
about seven hundred million. Some years ago Mrs. Gandhi was
turned out as prime minister reportedly in part because of her strong
initiative in family planning. Since then family planning in India has
had a low profile.

We learn that Mr. Chandra does custom work for his neighbors.
He has a plow, and with his bullocks he prepares the land for those
who have no draft power. His charge is forty rupees per acre, about
$4.50.

Other farmers, we find, have supplemental income. One with a
very small farm runs a tea shop in his village.

"Where do you get your irrigation water?" we ask Mr. Chandra.

"My neighbor has a well. I buy water from him. Wheat takes four
to five irrigations, which costs two hundred rupees." This comes to
about twenty-two dollars per acre.

Irrigation has taken a new and fascinating turn. There has long
been canal irrigation, put in by the British. But now the farmers have
discovered that there is a groundwater supply recharged annually with
outflow from the Himalayas, and that it can be reached by tube wells.
A tube well is a pipe, driven or drilled down to water, in contrast with a
dug well, put down by hand and lined with some kind of masonry.
Tube wells are cheaper and can go deeper. In 1961 there were
160,000 tubewells in India; in 1981 there were more than 8,000,000,
about half diesel and half electric. These wells are potent insurance
against drought and famine. But they do not provide complete protection. This year, with a weak monsoon, the water table has dropped.
Some wells have been deepened, and some fields have gone without
irrigation. We see a shallow well for family use, operated by a hand
pump, and another well sixty feet deep, that had been dug and lined.
We learn of driven and drilled wells that are quite deep, fitted with
four- or six-inch pipe, serving twenty acres.

Mr. Chandra's home in the village has tap water and electricity. He
has no radio. He has a buffalo for his family milk supply. The family
consumes most of the crops grown on the farm.

He estimates his total cash income from farm sales and custom work at about three thousand or four thousand rupees per year, or somewhere in the neighborhood of four hundred dollars, this for a family of three adults and five children.

"Have you made financial progress during the past ten years?"

"No," he says, apparently quite reconciled to this fact. But he has fed and educated his children and kept his land.

Perspective is needed here. Cash income is not a good measure of well-being for a family that lives largely outside the cash economy. In addition to cash income, the family has home and food supply. Education and health services are supplied by the government. The Overseas Development Council has worked out a quality-of-living index based on such measures as life span, educational attainment, and caloric intake. This measure is much superior to the computation of cash income. By the quality-of-living index, the Chandra family would still be poor, but relatively less so than by measuring cash alone.

Mr. Chandra, together with more than half of India's eighty-two million farmers, falls in the marginal farmer category, having less than 2.5 acres. These marginal farms account for 11 percent of India's farm area.

There are among the rural people many who are worse off than marginal farmers like Mr. Chandra. These are the landless laborers; they are dependent for their livelihood on what they can earn by employment, on farms or elsewhere. For them a new National Rural Employment Programme has been launched. But this has not been started in Haryana, where the people are better off than in most other Indian states.

"What would you do if you had a crop failure?" we ask.

"Find other work," is the answer. He says he can manage.

We learn that this farm has long been in the family. "Do you expect your sons to farm?"

"That is for them to decide. The children must get education. Then they can do whatever is best."

To the first-time visitor from the United States, a call on a marginal farmer like Mr. Chandra would be depressing. But we visited India in 1958, when many of the refugees from the partition had not yet been settled. And we visited in 1967, during the Bihar famine. Things are better now. The Green Revolution has lifted yields and transformed agriculture. The country no longer depends on food relief from the United States. We see more good farm equipment and fewer derelict cows. Electricity is coming in. Educational attainment has risen. Health service has improved. The caste system gradually gives way to the forces of modernization and change.

To the northwest is the Soviet Union. To the northeast is the People's Republic of China. These are, like India, enormous and populous countries. Russia and China have, each in its own fashion, chosen collective farming as the basis of agricultural development.

Indian farmers are traditionally individualistic, entrepreneurial, and family-oriented. Democratic socialism, Indian style, has endorsed an agriculture based on family farms.

India's farm population will continue to grow. There is no way the prospective increment of people can all find off-farm employment; the country cannot possibly industrialize that fast. The tiny farms will become smaller still.

Can India succeed with the choice she has made?

India's answer seems similar to that given by her citizen Mr. Chandra in response to our question regarding the prospect for his family. It is impossible to know for certain. India must be herself, do what she can, educate the children, and hope for the best. Like Mr. Chandra, India thinks she can manage.

Java

Adapt or Die

Bogor, August 29, 1982. This is Indonesia, an archipelago of 150 million people, the fifth most populous country in the world. There are some thirteen thousand islands lying astride the equator, spread over an east-west distance as great as that between New York and San Francisco. The central island, Java, is one of the most heavily peopled areas of the earth. Some of the other islands are thinly settled, having enormous areas of virgin forest. Eighty percent of Indonesia's inhabitants live in rural areas. Population is growing at a rate of 2 percent per year; two-thirds of the people are under thirty years old. If population growth continues at this rate, it will double in thirty-five years.

Feeding this population in the years ahead will be an enormous task. With what combination of farming systems will Indonesia try to meet this challenge?

The institution used by the Dutch colonizers to develop the country's agriculture was the estate, a big, privately owned farm, organized for profit, centrally directed, with labor, land, and management separately supplied. The estate is believed to have the efficiencies associated with large enterprises, called "economies of scale." The estate seems especially suited to the production of tropical crops for export, of which Indonesia has many: sugar, palm oil, copra, coffee, tea, and rubber.

The estate is known by several names. The Spanish variously call it an hacienda, a *finca*, or an *estancia*. The Portuguese call it a *fazenda*. In Australia it is known as a station. In the southern United States it was a plantation; along the Hudson River the Dutch of colonial times tried unsuccessfully to establish it as the patroon system. Its modern form is the corporate farm. The generic name is latifundia.

Advocates of the estate system maintain that not only is it efficient but also it provides a beachhead from which superior technology can penetrate and modernize the historical pattern. Traditionally it has been labor-intensive, providing much employment in countries where labor is abundant. It develops entrepreneurship and managerial skills which, it is said, will in time convert primitive agriculture to the superior system of the already developed countries.

Opponents of the system attack the estates on opposite grounds. These are the charges: It is operated in the interest of the imperial power, not with concern for the host country. Earnings are sent back to Europe. Labor is overworked and underpaid. Instead of lifting the general level of the rural areas, the estates become privileged enclaves. They exercise undue political influence, sometimes corrupting local officials. The emphasis is on crops for export; crops that help feed the people are neglected.

Nationalists hate the estate because it is foreign-based. Communists hate it because it is capitalistic. Its defenders love the estate not only for its alleged advantages but also for the enemies it has made.

In 1945 the Indonesians achieved independence from their Dutch overlords. How does the new nation look on the estate system, that legacy of colonialism? We put the question to Soewondo Kartokeosoemo, an officer of the Indonesian government, the man responsible for supervision of the state-owned plantations. He is in his forties, a native of Indonesia, holder of a master's degree from Texas Agricultural and Mechanical University, fluent in Indonesian, Dutch, and English. During colonial times his father was a tax collector for the Dutch government.

"We are pragmatists," he says. "We didn't throw out the estates. We transformed them to meet the needs of the country."

"How were they transformed?"

"We took over the Dutch estates and operated them as part of a government corporation. In effect, they are state farms. The estates owned by British, Belgians, and Americans we left as private enterprises, with conditional titles. As a matter of fact, we are inviting foreign capital into the country, to set up private estates in joint ventures with the Indonesian government."

"Can you explain these joint ventures?"

"We contract with the estates to develop new lands in the unsettled parts of Sumatra and Kalimantan. The estates know how to bring oil palm and rubber into production. Once we have a going operation, we can pass title to the settlers and have a group of smallholders. We use the processing facilities of the estate to serve nearby smallholders. It is a mutually helpful relationship. The estates have superior germ plasm for the tree crops; we see that this is shared with the small-holders. The extension service takes the science and technology of the estates and spreads them among the smallholders. By this pattern of service the estates earn the good will of the country. We have a mixed system."

But we learn in conversation with other Indonesians that while some of the smallholders respond to these overtures, others do not. Some of the smallholders think of the estates as we Americans might think of space creatures colonizing the United States.

"Is there an estate we could visit?"

There is. It is Gunung Mas (Gold Mountain), a tea plantation near Bogor, some seventy-five miles south and east of Djakarta. We ap-proach Gunung Mas by a winding mountain road. On either side are small terraced and irrigated farms producing rice, cassava, corn, and various fruits such as bananas, durian, papaya, pineapple, jackfruit, citrus, and lychee. Palm trees grow, singly and in clumps. Some rice fields are the size of our living room; some as big as a tennis court. The largest ones are up to an acre. Rice is in various stages, from newly planted to heading out. There are vegetables such as potato, cabbage, tomato, lettuce, eggplant, and carrots. People are everywhere, mostly young, healthy-looking and seemingly happy. Homes are small and low. Everything human seems miniaturized. We see chickens and goats, but almost no large animals, and no tractors.

Smallholders in Indonesia total about sixteen million units, five times as many as the number of farmers in the United States. Small-holder farms average about two and a half acres and occupy 86 percent of the land in farms; the estates have the other 14 percent. Average income per person on farms is the equivalent of about one hundred United States dollars and is only about 60 percent as high as in urban areas. The impression of wealth conveyed by downtown Djakarta is deceptive; in the rural areas cash incomes are low.

Gunung Mas is a tea estate of some twenty-three hundred steep and rolling acres. The administrative center is thirty-seven hundred feet above sea level. Altitude and topography of the estate are good for tea, which likes it moist and not too hot, with good air drainage. The mountain towers high above. Soil is volcanic, deep, loose, mellow, and reddish purple. Rainfall is 175 inches a year, five times as much as in Indiana, so that the soil is somewhat leached. There are about 220 rainy days annually, fairly evenly distributed. Despite the high rainfall and the steep slope there is relatively little erosion. The canopy of tea breaks the force of the rain, and the open volcanic soil takes up the water like a sponge.

The tea bushes are pruned to be flat-topped, at waist level, convenient for the pluckers. The mature bushes barely leave room for passage. Ordinarily the leaves are a bright glossy green, but at the time of our visit they are greyish brown with a film of dust from the volcano Mount Galunggung, which is erupting some sixty miles to the east. The leaves will have to be washed after picking.

There is a factory on the estate for drying, processing, grading, and packing the tea. The factory produces costly orthodox black tea for export, using a process that is almost a century old. This tea is classified as "good medium." Eighty percent of Indonesia's tea is of this general type and goes to export. The remainder is green tea, or scented tea, which is less costly and is preferred in the domestic market. The difference is in the manufacturing process. The tea is packed in boxes which resemble those tossed into Boston Harbor by the American patriots more than two hundred years ago.

The estate employs about sixteen hundred people, eleven hundred of whom live on the site, with their families, in three villages, each village having its own acreage to tend. In addition there are about five hundred casual workers who live off the estate, coming back and forth daily. Plucking is done on a rotation basis; each bush is visited once in about ten or eleven days. Traditionally it is a woman's job. The women wear broad-brimmed straw hats and carry sacks or large open-mouthed hampers. With knives they pluck the new tea leaves at the two- or three-leaf stage and pass them into the sack or hamper. It is piecework; they earn the equivalent of from one to one and a half United States dollars a day, somewhat more than they can earn from alternative employment. There is a limited amount of health service. There are schools for the children. The workers have a labor union. Management reports union leadership as "responsible people, not agitators like during the communist time." Working conditions, compensation, and fringe benefits are superior to those of the colonial era. We do not have an opportunity to talk with the workers. The assumption seems to be that the manager and his top people will tell us all we need to know.

Gunung Mas was established by the Dutch as a tea estate in 1910. It flourished until the Japanese invasion in 1942, when it was converted to cassava and other crops to meet critical food needs. Tea was replanted after the war. As they left, the Dutch sold the estate to Indonesians of Chinese parentage who in turn transferred it to the Indonesian government, for compensation. The amount of the compensation is not reported.

Resident manager of the estate is Pali Permana. He has three Djakarta-based supervisors, one for production, one for factory, and one for administrative affairs.

Production of Gunung Mas is twelve hundred pounds of tea per acre, somewhat below the average for state-owned plantations. Overage bushes and leached soil are the cause of this relatively poor performance. Replanting is under way.

On the whole, yields on the government-owned estates rate high as compared with other sectors of agriculture:

1979 Yields, Pounds per Acre

	State-owned plantations	Private plantations	Smallholders
Coffee	566	439	514
Tea	1,290	500	594

The country is anxious about its food supply. The estates have good technology and demonstrate superior performance. This is a major reason for keeping them.

The estates continue their interest in new genetic material and new methods. There are research stations for the export crops in various parts of the country.

To a newcomer the questions lack obvious answers. Does the estate have adverse social attributes, as has been claimed? Is it an outdated relic of colonialism? Or is it really a superior way of organizing agricultural resources, as some maintain?

The Indonesians, being pragmatists, avoid ideological answers to these questions. They neither embrace the estate in its old form nor reject it outright. They have undertaken to change this institution, to reshape it according to their needs.

Estates earn foreign exchange with which Indonesia can import food that it cannot efficiently produce, like wheat and soybeans. This will be increasingly important as the population grows and foreign exchange earnings from oil exports taper off.

Many years ago those huge creatures known as dinosaurs were the dominant form of animal life. Conditions changed, and the dinosaurs did not. So they became extinct. They did not adapt. So they died.

The estate was the dominant agricultural institution of the colonial powers. Conditions changed. Where the estates changed to meet these new conditions, they survive. Indonesia is a country in which this change is occurring.

The Indonesians see no merit in a monolithic agricultural system. To feed their people they plan to use any combination of systems that will produce results.

Malaysia

Clearing the Jungle

Bilut Valley, Pahang, November, 1962. Mohamed Ismail is a new farmer. That is, he never farmed before, nor did his parents. Until a year ago he was a fisherman from Trengganu, on the east coast of Malaysia. He was married, thirty-five years old, and the father of three. Like many fishermen he was poor; it was hard to provide for his family. He learned that the government of Malaysia was carving new farms out of the jungle in the interior of the country and that people could apply to become settlers on these new farms. What could he lose? He made application and by the luck of the draw found himself with his family on this new five thousand-acre farm in the Bilut Valley of central Malaysia, on the mainland.

The forest rises vertically one hundred feet or more on all sides of the farm. Monkeys chatter in the trees. Jungle birds sound their wild and raucous cries. Sometimes a python crawls out from the forest, to be killed and eaten for his tender flesh. Tigers lurk in the jungle, usually keeping to themselves. There are other creatures: elephant, rhino, tapir, wild pigs, crocodiles, and cobras. There is the *seladang*—a Malaysian gaur, a massive wild ox. Sixty percent of Malaysia's irregular surface is covered with dense rain forest. Seen from the air, it looks like a great green tufty rug thrown over the playthings left by a child on the living room floor.

There will be about five hundred families on this farm when settlement is complete. Many are already here. Mostly they are young, poor folk like Ismail. They live in simple homes, clustered around the farm headquarters.

Some fields have been cleared for several years, and some clearing is underway. In the forest are occasional giant trees, four or more feet in diameter, towering high above their neighbors. There are as many as one hundred different species on a single acre, some of them hardwoods so heavy they will not float. In Kuala Lumpur, the capital, clear mahogany is used for scaffolding and for concrete forms.

The trees are felled with power saws. They lie every which way, piled on top of one another, crisscrossed like so many giant jackstraws. Marketing the timber does not pay. Roads are poor, distances are great, and prices are low. So the trees are burned as were the trees on the American frontier 150 years ago. During the dry season, which is the time of our visit, fires are started, visible from the air as columns of whitish grey smoke rising thousands of feet.

The burn is never complete. When it dies down, the small trees and brush have been consumed. But the big logs lie half-burned and blackened, the jagged stumps still thrusting up, a scene of utter desolation.

Now come the bulldozers. They nudge the fallen giants aside in accordance with a planting plan, based on the contour, which is definitely rolling. Place is made for setting out young oil palm. These are a new, high-yielding variety, developed in the research station and grown in the nursery. Only about a foot or two high when set out, the young trees can hardly be seen amid what remains of the forest. Mohamed and others of his crew set out these trees, applying the prescribed amount of fertilizer, following the instructions of the foreman.

A legume is seeded among the half-burned logs between the rows of young palms. In the chocolate brown soil, high in humus, and with abundant rain, the seeds sprout quickly, and vines soon creep over the tangle of logs. Mohamed and his crew fight the crawling green away from the young palm trees, which would otherwise be smothered.

The legume carpet soon covers everything but the young palm. Beneath this carpet, temperature and humidity are such that the half-burned logs decay. So the palms shoot up, the logs rot down, and the green carpet spreads all about. In five or ten years what was a scene of havoc becomes a place of beauty. The legume carpet not only helps rot the logs; it also checks erosion, chokes out weeds, reduces soil temperature, and fixes nitrogen for the young palm.

In some places, efforts to bring tropical forest land into crop production have been disastrous. With the leafy canopy destroyed, the tropical sun raises soil temperatures, burning up soil organic matter which, with the trees gone, is not replenished. The soil in such cases quickly loses its fertility and its texture. The rain, with nothing to break its fall, hammers into hardness such soil as it does not wash away.

Not so on this farm in Bilut Valley. When they grow up, the palms will provide a cover, as did the original forest. Meanwhile the legume breaks the force of sun and rain.

When Mohamed is not planting, he is building roads, erecting farm structures, tending young palm in the nursery, or felling forest. All the skills are new to him. He has nothing to unlearn. He is more teachable than someone whose experience in traditional agriculture leads him to think he knows it all. That was a factor in Mohamed's selection. The leaders of the American poultry industry demonstrated the value of an inexperienced worker when they transformed egg and broiler production. Greatest success was with people who had never tended a chicken.

Mohamed and his family have use of a small tract of land. Here they produce food for themselves. There is upland rice, rainfed, which does fairly well with one hundred inches of precipitation per year. There are pineapples, banana, papaya, and other fruits. Manioc is in production, and young coconut trees are starting. Chickens and goats, dear to the Malays, provide companionship as well as food.

The older of Mohamed's children are in school. Health services are available to all. Radio provides contact with the outside world.

The oil palm likes abundant rain, bright sunshine, and warm weather, all of which Malaysia has. It likes rich soil, which is found in certain spots but not generally in the country. Interestingly, both of Malaysia's two big export crops, rubber and palm oil, were domesticated in the twentieth century. Each was brought in from afar, the rubber originating in the Amazon basin and the oil palm in West Africa.

A few years after planting the oil palm begins to produce clusters of fleshy date-sized fruits, deep red in color, each with a stony kernel. These bunches are cut and carried to the gathering road, where they are picked up and hauled to the processing plant. Oil palm is a labor-intensive crop, well suited to a country with an abundance of human resources.

To be efficient, a processing plant should be of such a size as to handle at least five thousand acres of palm. This dictates the farm acreage.

When processed, the fruits yield a thick oil, dark orange, having a pleasant odor. The oil is good for several uses, food being the major one. It is used for shortening, margarine, and salad and cooking oil, as well as in soaps and lotions. The kernels are also processed; they yield palm kernel oil, used chiefly for margarine. An acre of mature oil palm will produce ten times as much oil as an acre of soybeans. Trees will yield well for many years.

Technology is excellent. Research work has not only produced new, high-yielding varieties; it has also pointed the way to good cultural practices. Palm leaves are analyzed chemically, making possible the prescription of the appropriate amount and kind of fertilizer.

For many years Malaysia's export crop was rubber. But with synthetic rubber coming on, need was felt for diversification. Palm oil was chosen for development. Not only is the oil palm tree well adapted to Malaysia's soil, climate, and labor supply; it has a growing market. The world population increases, and per capita income rises. As poor people experience rising incomes, they improve their diets. One of the foods they increase most readily is food oil; it provides energy, improves palatability, and is reasonable in cost. The economist would say the vegetable oils have high income elasticity in the Third World. Vegetable oils like palm oil cost much less per pound than do the animal fats like butter, lard, and tallow. Nutritionists rate them higher than the animal fats.

The peninsula of mainland Malaysia is a long, fingerlike projection dangling down from Asia pointing toward Indonesia. The population is concentrated near the coast. The native people are Malays, dark-skinned and small, similar in race, language, and culture to the Filipinos and the Indonesians. The Malays traditionally have lived in rural clusters called kampongs, in houses raised on stilts. They produce their own food; the staples have long been rice and fish.

Chinese were brought into Malaysia to work the tin mines, and people from south India were brought over to tap rubber trees. Deep in the forest, keeping mostly to themselves, are groups of aborigines, the Stone Age people found in the remote areas of Southeast Asia.

The Malays are the most numerous, constituting about half of this mixed population. Mohamed is a Malay. His religion is Islam, but there are in his faith some vestiges of the animism which prevailed generally prior to the arrival of the Islamic faith many centuries ago. There are other settlements for the Chinese and the Indians.

Malaysia has a complex history, which includes Dutch, Portuguese, and English periods. It became independent in 1957. Of the
various Third World countries, Malaysia is one of the most rapidly
developing.

With the death rate now generally under control, population increases at a rapid rate. Despite substantial economic development,
new industrial jobs cannot absorb all the new job-seekers. The majority of the people are engaged in some form of agriculture. There is
opportunity to expand agriculture; only about half of the potentially
tillable land is in production. Expanding agriculture would provide
jobs, increase export earnings, diversify the economy, and broaden
the country's political, economic, and social base. In addition it
would diminish the likelihood of civil disturbance. The country's
leaders remember how the Communists exploited the poverty and
joblessness that followed World War II. An enormous effort was
required to put down this uprising during the emergency. Malaysia is
one of the few countries that overcame a strong guerrilla movement.

Soil surveys are undertaken to locate the areas where agriculture is
most likely to succeed. This is crucial. If the soil is infertile, failure is
likely no matter how favorable all other factors may be. This was
found in the American experience. Moving west with oxen and covered wagon, at some point the settler said "Whoa!" Whether he spoke
this word on fertile or infertile ground determined in large part
whether his descendants turned out rich or poor.

Legally the undeveloped land in Malaysia is owned by the ruler of
each particular state or province. This ruler, the sultan, retains ownership unless he decides to "alienate" the land. That is, he may transfer
ownership to another party. Thus before the Federal Land Development Administration can proceed, it must persuade the ruler of the
state to alienate a tract of land, a delicate negotiating matter. The
rulers are reluctant to part with what they consider their patrimony.
The landholding institutions of Malaysia are an amalgam of traditional Malay practice and Islamic law. The rubber plantations have
tried with limited success to impose a set of Western proprietary ideas
on this complex landholding pattern.

At the time of our visit the farming system on the new settlement in
the Bilut Valley is like that on the big rubber plantations: the farm is
large; the workers labor in crews, for pay, under foremen; they have
living quarters as family units; and they receive social services. In

some respects the system is reminiscent of a Russian state farm. The settlers are told that they will in time pay for and own individual tracts, in the tradition of Malay smallholders. We learn later that this does indeed happen. Many countries are noteworthy for making plans. The Malaysians are noteworthy for following the plans they make.

Enormous investment is necessary to bring in one of these farms. There are the costs of land acquisition, clearing, and planting. Roads must be built, structures erected, and a processing plant put up. In addition, living costs of the settlers must be met during the years of waiting until the palm trees come into bearing.

There is always a wait between the launching of a land settlement venture and the time when the enterprise becomes self-supporting. In the United States it was said that one generation settled the land, a second generation improved it, and real net benefits did not come until the third generation. A people must have a strong commitment to the future to undertake land settlement. Most individuals today seem disinclined to postpone rewards for so long a time. Consequently land settlement schemes are heavily subsidized.

But there are some who undertake land settlement on their own, despite the difficulties. There are squatters in Malaysia, just as there were on the American frontier. A squatter will find a spot, reasonably accessible to a road, and put up a shelter. He will girdle some trees to let in the sunlight. There in the open area he will grow some rice, manioc, and pineapple, with perhaps some bananas and papaya. He will keep a few chickens and perhaps a goat. He is trying to emulate, as well as he can, the traditional Malay subsistence form of livelihood. He will become self-reliant at a very low level of living in a fairly short time.

This form of settlement generates problems. First of all, it is illegal, though the laws against it seem unenforceable. And the soil may turn out to be unproductive, so that the site is soon abandoned. Erosion may bring irreparable damage. It is almost impossible to supply school, health, and other public services to these remote and scattered homes. Nevertheless, we learn that more families are settling this way than on the Federal Land Development projects. In the long history of agriculture, individual settlement, undertaken without central plan, has been the dominant form.

Why not let private enterprise develop new oil palm plantations, as was done in the past with rubber? Government is wary of giving full

support to the plantation system. Private capital, noting the expropriation of plantations in many countries, is cautious about heavy new
investment.

However all this may be, Mohamed is hopeful. He now has a good
food supply for his family and better social services than they have ever
before known. He has a heavy discount for the future.

In some countries, hunger for agricultural land takes the form of
conquest or expropriation. In Malaysia it is expressed in agricultural
development. Instead of dividing up the already developed farms, this
country brings in new land.

Viewing this conversion of raw wilderness into farmland, one is led
to the fact that every acre now in production went through some such
process. The economist David Ricardo said in the early nineteenth
century that the supply of farmland was fixed. It was a useful assumption, making possible some titillating debates in economic theory. But
it was wrong. There is opportunity to develop much more farmland.
Globally, the acreage of tilled land could be doubled.

This is under way. In the Netherlands, farmland is reclaimed from
the sea. In the Middle East, it is redeemed from the desert. In the
savanna area of Brazil farmland is created from brush and grassland.
In Bangladesh, swampland is being made into farms. In Malaysia the
tropical forest gives way to farmland, and in Finland the conifer
timberland likewise recedes.

Some people believe that while this may be an economic necessity,
it is an ecological disaster. This need not be the case, as testified by the
Bilut Valley settlement.

Portugal

The Counterrevolutionists

Évora, July 16, 1982. "The land is mine. I own it. They took it. I've got it back, and I'm going to keep it!" These words, with an edge on them, come from João da Silveira, usually soft-spoken. Silveira is a farmer from southeastern Portugal, the breadbasket of that troubled country. He and his sister hold three thousand acres of rolling upland, dotted with pine, cork oak, and eucalyptus. The broad landscape sparkles under the July sun. Red Alentejo cattle, native to this area, graze in the distance. Silveira's men pick up straw bales to be used for winter feed, the last stage of the oat harvest. On the horizon is Évora, a town of forty-five thousand, an ancient walled city going back to Roman times. There is an old aqueduct and a temple to Diana. The city has now spilled out beyond the walls. The land here has been farmed for at least two thousand years; no one knows how much longer.

Portugal has a two-sector agriculture. Here in the Alentejo, comprising about one-third of this Maine-sized country, most of the land is in large farms and has long been so. These farms are fairly modern and provide their owners with what is on the whole a good livelihood. In the Alentejo district before the 1974 revolution, the larger farms, though they constituted less than 2 percent of all farms, had 57 percent of the land. In Portugal as a whole almost 40 percent of the farms had less than two and a half acres apiece. These were mostly subsistence farms, largely outside the exchange economy. Living levels on these tiny farms were very low. In 1968, 44 percent of the farm people were illiterate.

Here in the Alentejo are the remnants of the feudal system which the Old World exported to the New some four hundred years ago.

From 1926 to 1974, Portugal was ruled by a dictator. In the 1970s things began to come unglued. Its African empire crumbled under left-wing attack. Mozambique and Angola were lost. On April 25, 1974, left-wing leaders threw out the existing Portuguese government and took power.

When the revolution came, the common people invaded many of the big farms in the Alentejo district and took possession. One day, not long after the coup, Silveira was confronted on his farm by a group of men consisting of his own hired workers, other landless laborers, and some adventurers. They told him they were taking over; the farm was now theirs. He had thirty days to remove his personal possessions.

"Why do you do this?" he asked.

"The leaders of our Farmworkers Union tell us we must. If we do not, others will take the farm, and we will have neither job nor home."

"Will you sign a paper stating what you have told me?"

They would. Those who could write their names signed the paper.

Silveira, then in his middle thirties, took his wife and young child and moved to Lisbon, where he lived with his wife's people for a year and a half. He took with him his heirlooms and such valuables as could readily be moved. Silveira himself knew terror and violence; he had fought in the Angolan War. He could cope. But it was an ordeal for his wife.

The Silveira farm had been church property, a convent, in the 1600s. The Silveiras acquired it in 1810. João is the seventh generation of his family to possess the land. His young son was to be the eighth. That family succession was disrupted.

The revolutionaries who took over the farm were organized into a cooperative after the fashion of the kolkholzes of the Soviet Union.

In the three districts of the Alentejo region, nearly two and a half million acres of land changed hands, 54 percent of the total land in the area.

The spontaneous invasion of farmland was legalized, modified, and corrected with the land law of 1977. As of the end of 1980, about 130,000 acres, or 5 percent of what had been confiscated, had been given to small farmers in tracts that averaged about 50 acres each. About 1,100,000 acres were returned to the former owners, with a variable ceiling of 1,250 acres for each individual. Actual size was based on a point system that depended on the type of land. About half of the land that was expropriated remains in the hands of 289 cooperatives.

The revolutionists, while in power, rewrote the constitution in socialistic terms. Then, in 1976, true to their democratic rhetoric, they held an election. They won only 16 percent of the vote. So a coalition came to power, the Democratic Alliance (AD), a broad-based group held together by mutual opposition to the revolutionaries. There is now an anomolous situation: a center-right government operating under a left-wing constitution. A house divided. Farmers like Silveira don't know how strong is—or will be—the legal case and / or the political power structure on which their landholding is based.

When the revolutionists took over the Silveira property, they neglected the cattle enterprise, and the herd deteriorated. Nationwide, agricultural production, already low, declined further. Wheat production fell 40 percent below prerevolutionary levels.

When the tide began to turn against the revolutionists, Silveira went back to his farm. He was one of the fortunate (wise? resourceful? influential?) ones to whom the land was returned. He brought back his personal possessions, mechanized the farm, cut his labor force from eighteen to six, and reestablished his cattle-breeding program. He did not follow through on earlier plans to build three small dams for irrigation; the future was too uncertain.

Silveira is of average height, clean-shaven and erect. He speaks good English. His home is the convent of four hundred years ago, solid masonry, one floor, low-roofed, heated with huge fireplaces. There is a beautiful altar, originally the worship center of the convent, kept with care since the 1600s. Down the hill from the mansion and some distance from it are the homes of his workers, one-floor row houses, of masonry, providing an abundance of space for the much-reduced labor force. On the hill stands the base of the windmill which once ground grain for the enormous labor force that formerly worked this farm; one can almost picture Don Quixote tilting with it. In former times the farm was a unit in itself, providing for almost all its needs. There is a huge barn with an open stall, once used to house the forty bullocks that provided draft power. There are six new horse stalls; Silveira loves horses and breeds them, for riding and driving, not for draft. He has trophy cups won with his animals. Listening to this man, one has a feeling for the depth and breadth of the tradition, the history, the sense of family, and the commitment to proprietorship that are the essence of this agricultural system.

* * *

Present during the visit to Silveira's farm is his friend, Jose Maria Queiroga, also a farmer from the Évora region. He is national president of a center-right farm organization, Conferação das Agricultores de Portugal, or CAP. Queiroga is a hale and hearty man in his upper forties, the father of four. His face is tanned, his hair is cropped short, and he has a great moustache. He wears a chamois-skin jacket, in natural color. He has a gracious manner and excellent presence. He holds a degree in agronomy from Lisbon. Like Silveira, he speaks English. Queiroga operates about 650 acres, half of which are grazing land and half grain. His grandfather farmed these same 650 acres, but his father lost 400 of them. This land Queiroga has regained, as a matter of family honor. Cork oak and olive trees are scattered over his farm; his five hundred sheep and fifty cows graze under the trees. He draws a sketch showing how, every nine years, the cork is peeled from the trunk and limbs of the tree, after which it renews itself. He farms and grazes his animals under these trees and around the rock outcrops that appear here and there. The soil is sandy and thin, underlaid with tight subsoil so that interior drainage is poor. Rainfall, which averages only about twenty-three inches per year, comes mostly in the winter, when it is not needed. June, July, and August, normally the growing months, are very dry. Fertilizer doesn't help much; moisture is needed to convert the elements to nutrient forms that the plant can use, and moisture is short during the summer. Grain yields in Portugal in recent years have averaged only one-fifth or one-fourth as great as in the European Common Market, and only half as great as in Spain.

Queiroga's farm was one of those invaded during the 1974 takeover. During the disturbance, Queiroga worked in a fertilizer factory.

"What is needed to improve agricultural performance in Portugal?"

Queiroga is quick with his answer: "Stable government. An end to uncertainty about land policy, so farm people will invest. Roads.

Drainage. Irrigation. Credit. Less bureaucracy. Rural electricity and phone service." He has waited seven years for a phone and still doesn't have one.

Does he have confidence in the new school of agriculture and the agricultural experiment station being set up in Évora with the help of the Americans? He waits to see. "Maybe, if they work on practical problems."

Queiroga's farm organization, the CAP, is a federation of many district organizations. It was founded after the revolution; its purpose is to protect landholders against the land reformers. Greatest strength is in the south and east, where the land reform was most active.

"What is your position on Portugal's proposed entry into the European Common Market?"

Queiroga flashes a quick smile. CAP pushed for entry. The reasoning was that in order to join the Common Market, Portugal would have to have an agriculture compatible with that of the other members. The form of agriculture favored by the revolutionists could not relate satisfactorily to the entrepreneurial agriculture of the others. The CAP won the campaign; Portugal intends to enter. Now that this is the prospect, the CAP is trying to learn what it will mean to Portuguese farmers.

These two men are counterrevolutionaries. The principle, verified by much of history, is that a revolution, carried on with zeal and verve, usually goes beyond what the public processes will accept, long-term. The excesses set in motion a reactionary force that pushes back toward (but not all the way to) the original position. So it was, the historians say, with the American Revolution as well as the French, the Russian, and the Chinese. So it is with the Portuguese. And so it will likely be with the Salvadoran. One should not easily dismiss a man's tenacity in fighting for land that he considers his own.

As we leave Silveira's home I notice several guns in a rack standing against the wall. "Do you hunt? Is there game here?"

There is a pause. The two men exchange glances. Then Silveira says evenly, "There is no game here." Another silence. We speak of other things, say our goodbyes, and depart for Lisbon.

Taiwan

The Part-Time Farmers

Taipei, October 2, 1982. There once was a clean-cut dichotomy, farm on the one hand and nonfarm on the other. This line of demarcation portioned out the populace not only on the basis of vocation but also by social status and living habits.

But that is changing. Farmers have been adding off-farm jobs to their farming operations. On this trip we see the trend in almost every country we visit. In the United States the off-farm earnings of farm people now exceed their incomes from farming.

The reasons for this trend are varied, subtle, and powerful:

Population increases, but the acreage of land in farms does not increase proportionately.

Divided through inheritance, some farms become too small to provide adequate income.

Mechanization of farming increases, and hence there is a reduction in the number of farm jobs.

Industrialization of the economy goes forward, and so more nonfarm jobs are available.

Increasing education on the part of farm people makes for greater willingness and ability to do off-farm work.

Better transportation makes it easier to take a job away from home.

Love for the land and the rural tradition keep a family farm-based even though much of the income is of nonfarm origin.

The result is part-time farming. People find part-time farming difficult to conceptualize. It straddles the border between the two historic categories, farm and nonfarm, and the mind resists the violation of so important a boundary.

For many years a debate was waged. Some said part-time farming was the midway stage for a man entering agriculture. Others said it was an intermediate point in leaving the farm. There was reluctance to acknowledge part-time farming for what it in fact had become: an established occupational form.

The trend toward part-time farming has gone farther in Taiwan than about anywhere else. Population is dense and growing. Farms are small and modernizing. Taiwan is industrializing rapidly. Education and transportation are advancing. Distances to off-farm jobs are short. And the Chinese have a traditional love for the rural life that leads them to keep one foot on the land.

In Taiwan a full-time farmer is defined as one who receives at least four-fifths of his income from farming. Only one farmer in eight falls into this category. Part-time farmers constitute the overwhelming share of Taiwan's agricultural producers.

So we go to Taiwan to see what the agricultural economist would call "part-time farming in a developing market economy." What is happening on this island may have portent for other densely populated industrializing market-oriented Third World countries. There are a number of such.

Taiwan is an island about 230 miles long and 90 miles wide. It lies astride the Tropic of Cancer, 90 miles off the Asian coast. Its original name "Formosa" ("beautiful island") was conferred by the Portuguese. Thickly forested mountains run from north to south, taking up two-thirds of the area. An apron of gently rolling hills and level land borders the mountains on the west making up the main agricultural area. Rainfall averages more than one hundred inches a year. Short, swift rivers have cut gorges in the upland and spill their torrents into the lowlands after every heavy rain. Dams impound this flow, watering the terraced land on the slopes and the level fields of the lower river valleys, supporting a highly intensive and remarkably modern agriculture.

Rice is the mainstay, two crops a year. Sugar cane, sweet potatoes, asparagus, mushrooms, bananas, pineapples, and tea are grown. Chinese cabbage and other intensive vegetables are everywhere about. Hogs and poultry are the main livestock enterprises.

Twenty-eight percent of the people are farmers. The diet is similar to that of south China, their ancestral home. Rice is the staple grain, sweet potatoes the major root crop, Chinese cabbage the ever-present vegetable, bananas and pineapples the main fruits. When there is meat, it is likely to be pork. Duck, chicken, and fish are additional preferred foods, as are eggs. The people seem to thrive on this diet.

The island has a troubled history. Aborigines, similar to those in other parts of Southeast Asia, were the first inhabitants, and 150,000 of them continue to live in the mountains. Some Chinese came from the mainland as early as the 500s, but large settlements did not begin until the 1600s. The Dutch came early in the colonial period and were driven out by the Chinese in 1661.

In 1895 Japan gained control and developed the agriculture of the island. In 1945 the Japanese were thrown out. In 1949 President Chiang Kai-Shek moved his government to Taiwan. A million and a half civilians and troops came with him.

Under the Japanese the land had been in large holdings and was rented to the farm operators. Chiang Kai-Shek launched a land reform, transferring land held by absentee landlords to tenants. This change greatly reduced disparity of income and ushered in a prosperous time for farm people.

After Chiang's arrival, industrialization expanded with a surge that amazed the world. Both heavy and light industry leaped forward. Cement, fertilizer, plywood, plastics, television sets, and other consumer goods poured into the domestic economy and onto world markets. Job opportunities multiplied. The Chinese Nationalist Government, which had fought the Communists, was intent on proving that entrepreneurial economics was more productive than the collective system which they had opposed. Thus to the natural incentives of the competitive system was added the zeal of a dedicated people. The result was astounding.

Meanwhile, farmers were running into difficulty. The newly carved-out farms had been of reasonable size when the work was done by hand. But with the coming of tractors, herbicides, and harvesting machinery, many of the farms were too small. Division among heirs was a problem. The land reform, which had been a great blessing when it was instituted, had, with the passage of time, come to place a ceiling over agricultural opportunity. Farmers saw the urban sector modernizing and saw urban incomes rising. Unless they made changes, they were held to hand labor, small-scale operations, and incomes substantially below those of their urban cousins. The small farms could not effectively use the growing farm labor supply. What to do?

There were several options. One was to leave the farm, go to the city, and take an urban job. Some did so.

For those who elected to continue in agriculture, various alternatives were available. A farmer could stay small, continue with the old methods, and accept the limited income associated with traditional agriculture. A number took this route.

Still another option, which proved disappointing, was the course followed in the United States, to buy up a number of farms, consolidate them, and make a unit large enough to use modern methods and bring in a larger income. But this was not feasible for many people in Taiwan. We found land priced at thirty thousand dollars an acre, ten times as high as the price of equally productive land in the United States. Why is land so high? The Chinese have had experience with hunger and inflation and consider land a safeguard against these dangers. Land ownership carries enormous prestige. All of this prices land far above its income-earning capability and makes it extremely difficult to buy land at the going price and make a profit by farming it. The American idea that the price of farmland should reflect its capitalized ability to earn income misses the mark by several orders of magnitude in Taiwan.

Yet another option was to intensify operations on the limited acreage. Growing mushrooms, raising hogs in confinement, and producing eggs and broilers are very intensive and can bring in substantial income from limited acreage. The same is true of multiple-cropping vegetables. A number of farmers chose this route.

The option chosen by most farmers was to add off-farm income. This also took several forms. The one with which we were familiar was to add an off-farm job (driving a truck, doing construction work, taking factory employment) while continuing to live in the rural area and operate the farm. Many took this route.

But we found two institutional innovations that intrigued us, and we focused on these. One was custom farming. The other was joint farming. First, custom farming.

Tsai Yao Kun is a young farmer from Tsing Shui Park Township, Taichung Prefecture, in the rice-growing area on the coastal plain of western Taiwan. He has had six years of schooling. He is not only a farmer; he is also an agribusiness man. His services are used by eight hundred farm families. At the peak of the season he hires sixty people. He sells rice seedlings, does custom farming, and supplies farm chemicals. He is dark and lithe, a handsome man with curly hair, forty-three years old and the father of five children. He is a veteran of the 4-H Club, a cooperator with the extension service, and a supporter of the farmers' association. He has been in this business for about seven years, the first of which were hard going. But now he prospers, living comfortably in a large new house.

How does a young farmer like Mr. Tsai, beginning with only two and a half acres of land, succeed so quickly and so greatly?

Mr. Tsai began by growing and selling rice seedlings. The production of rice seedlings is one of the most technical and labor-intensive operations. Part-time farmers have inadequate skills and limited time for this demanding task. Mr. Tsai grew his plants in shallow flats, in a medium of earth and rice hulls, so that the roots could be separated with minimal injury. He tried several disappointing varieties before finding a rice that was adapted. He learned the proper scheduling, based on season, availability of irrigation water, and rate of growth.

Once he had succeeded in producing and selling seedings, some of his part-time customers asked him to do the transplanting as well. This he agreed to do, for a fee. He used a four-row transplanter, powered by a walking tractor. The rice needed spraying for weed control and to check plant diseases. This could best be done with pressure sprayers, difficult for a small operator to buy or operate, particularly if he worked all day on some off-farm job. So Tsai took over the spraying task and with it the supplying of farm chemicals. Then the harvest: Tsai had small combines, which permitted rapid reaping. Some of the part-time farmers wanted and paid for harvesting services. Gradually the operation grew until Tsai had taken over full operation on many of these farms.

The Taiwan Extension Service, which is among the more alert of such systems of the world, saw the agricultural development potential of this system. If they can get a superior variety of rice and persuade Mr. Tsai to use it, it quickly goes into production on eight hundred farms. This is the case with Tainung 67, a new short-stemmed, high-yielding rice developed by the Taiwan Agricultural Research Institute. Similarly with tillage practices and agricultural chemicals. Compare this with the separate education of eight hundred individual families, most of whom have their major income and interest elsewhere. Enormous leverage for improved agricultural practices is provided by custom farming, and this through voluntary entrepreneurial methods. The government puts in some subsidy to speed up adoption of the new practices.

Is Mr. Tsai likely to acquire monopoly power in the custom business? Little danger, say the government officials. The fees he can charge are subject to government oversight. The government can withhold subsidies if Mr. Tsai should overcharge his customers.

Will Mr. Tsai buy up some of these farms and get into what we Americans call a direct operation? Not likely; the price of land is prohibitive.

We visit one of Mr. Tsai's farmer clients. This turns out to be the young man's uncle, Tsai Yu Geng, seventy-one years old and retired. The elder Mr. Tsai owns about two acres of land. He has four sons and three married daughters. None of the children is interested in farming; all the second generation families are reliant on off-farm income. So the elder Mr. Tsai turns the farm over to his nephew for full operation, for a fee.

How will the farm be passed to the next generation? It will be kept intact; the four sons will hold equal shares of ownership, and it will be custom-operated. Will it be sold? They seem not even to have considered the question.

We visit with the elder Mr. Tsai. He has spent all of his seventy-one years on this farm. He recalled the Japanese period, before World War II, when the farmers were told what they had to do. Rents were very high; Mr. Tsai says little was left for him after paying the rent. He remembers the terrible war years, when food was requisitioned by the Japanese. Some of the farmers buried rice in the cane fields for their own use and, if found out, were severely punished. He remembers the American bombers; the most impressive fact was that they knocked out the fertilizer factory, reducing the food supply.

Then, in 1949, Chiang Kai-Shek came with his Nationalist Government. To Mr. Tsai the land reform was the major event of that time. His father had owned half an acre of land. With the land reform Mr. Tsai bought an additional acre and a half from the government, land the government had acquired from the Japanese. The cost was modest, the value of two and a half times the normal rice yield. The land thus acquired forms the basis of Mr. Tsai's present holding.

Mr. Tsai is enjoying his late years. Now retired, he feels no responsibility, keeps no records. When we ask a question about the farm, his nephew works out the answer with his pocket calculator.

A number of the old man's children and their families live near him, seemingly in comfortable circumstances. He has thirteen grand-children, several of whom we see. Though Mr. Tsai has no schooling, one of his grandsons is in college.

Custom farming is familiar to us in the United States, but here in Taiwan it seems to have taken a new direction. It seems that a system is developing in which ownership, labor, and management are almost entirely separate, with title to land widely dispersed among people who no longer farm. The system resembles stock ownership of a corporation.

Something like this has happened in northern France. The Napoleonic Code requiring equal division of the estate among the several heirs has resulted in landholdings so small as to preclude efficient operation. So entrepreneurial farmers solicit those who own fragments of land and for a fee obtain rights of operation. They consol-idate the lands to which these production rights pertain and run efficient, large-scale farms.

We next look at joint farming, which is entirely new to us.

We go to Wai Pu Township some miles distant, also a rice area. There are here 280 farm families who together have 750 acres of riceland. Many of these families rely heavily on off-farm income for their livelihood. Here no one has taken the lead, as has Mr. Tsai in the Tsing Shui area, to provide custom farming services. So the people, on their own, have invented a new institution, joint farming.

Headquarters of the joint-farming operation is a set of new build-ings in the midst of a rice field. We gather in a council room. One whole wall is given to an altar honoring the ancestors. There is a strong Buddhist tradition here, and the name of Confucius is held in high esteem. We drink ceremonial tea as the interview proceeds.

Joint farming is described to us by three gentlemen: Hong Tsing Piao, leader of the joint farming operation; Hong Tsing Tien, farmer owner of the rice nursery that serves the group; and Chiang Moo Ken, in charge of the credit department of the Wai Pu Farmers' Associa-tion.

Translation is ably provided by Lin Shih Tung, chief of the Inter-national Cooperation and Information Division of the Council for Agricultural Planning and Development, our host and guide.

The men outline the system, which, we later learn, was developed in Japan at about the same time. The 280 farm families are divided into seven groups of about 40 families each. In each group, about eight people operate the rice acreage on behalf of the various owners. The amount of service and the charge therefor are negotiated between the landowner (part-time farmer) and the operating man (full-time farmer). Some of the landowners elect to do a fair share of the farm work themselves; many others contract out virtually the whole operation. The phrase *part-time farmer* is in this case a sort of euphemism; most of the owners do little or no farming. The project overall is in charge of a team of three: a project leader, a comptroller, and a secretary. The idea was developed by the farmers themselves; the institutional format is the product of a government unit, the CAPD (Council for Agricultural Planning and Development).

Thus eight men essentially do the farming on forty ownership units. Modern farming methods are used; the aggregated units are of a size to permit what the economist calls "efficiencies of scale." We learn that tillage is 100 percent mechanized, as is spraying. Planting is 90 percent mechanized. Harvest is by combine; all rice is thus harvested except some that is lodged. The technology is excellent. Yields are good.

The extension people work with these joint farming groups as they do with the custom farmers. Some subsidy is supplied. Fields are consolidated where this is possible, to accommodate mechanization. Field roads are widened to permit passage of farm machines. The government supports the program by helping find off-farm employment for prospective members. Modest subsidies are paid to part-time farmers for joining the program and to the full-time farmers for taking over responsibility of operation.

Almost all the rice farmers in the area participate in the program, which is voluntary. The ones who elect to stay out are the small operators with no off-farm jobs and very little income to pay for services.

By these innovations the government and the people of Taiwan are making the adjustments called for by an increasing population, a fixed and severely limited land supply, a modernizing agricultural system, and a rapid industrial expansion. They are doing this with methods that are voluntary and entrepreneurial. They are splicing farm and nonfarm income.

Taiwan and mainland China have similar endowments: limited land and many people to farm it. In neither country is there enough

land per farmer readily to accommodate mechanization and modernization. Mainland China addresses the problem by centralized decision-making, putting huge units of land together into communes and
developing light industry within the communes. Taiwan uses voluntary methods: part-time farming supported by the institutional innovations custom farming and joint farming. The slogan in Taiwan is to
help people "leave the farm without leaving the village."

Very likely the potential of these new institutions is greater for rice
than for most other farm products, and is greater in the area we see
than in other parts of the island.

How is it that Taiwan has modernized its agriculture more rapidly
than most other parts of the world and has developed the institutions to
accommodate change? Our host, Mr. Lin, who holds an advanced
degree in sociology, has an explanation. The Chinese of Taiwan, he
says, are a "change-prone people." They settled this island from the
mainland centuries ago. As was the case with the western movement
in the United States, "the timid never started, and only the strong
arrived." Adaptation to a new setting required willingness to change.
The early Chinese had to cope with the Dutch, whom they evicted in
the seventeenth century. The Japanese occupation required further
adaptation. The restoration beginning in 1945 was yet another im-
posed change. The coming of the Nationalist Army in 1949 required
still more change. Throughout, the people of Taiwan have retained
their competitive entrepreneurial attitudes. The result is an economy
that is, on the whole, voluntary, open, and innovative. It is resilient
rather than either fluid or brittle. Mr. Tsai, a man of both tradition
and enterprise, characterizes this fruitful ambivalence.

As we leave, we ask what the future holds. There is a long discussion among the three who lead the joint-farming operation. They
agree on an answer: Part-time farming will increase. Joint farming and
custom farming will continue. The part-time farmers will continue to
own most of the land.

What have we seen in this part-time farming with its custom farming and joint farming? A glimpse of the future? An aberration? More
likely, promising innovations not yet ready for conclusive appraisal.

What is there here that other countries might emulate? It is clear to
us that Taiwan's farming model can succeed only in a country that is
industrializing rapidly and has an entrepreneurial economy.

However this all may be, the agriculture of Taiwan is in the process of losing its uniqueness. The Great Wall between farm and nonfarm is being breached. The farmers of Taiwan are entering the mainstream of economic, social, and political life. Skill will be required to navigate these new waters. But the Chinese on Taiwan are open to change if it is for a better life.

The Union of
Soviet Socialist Republics

Kalinin Farm

Belorussia, August 29, 1970. It seems a strange way to visit a farm, but
for us it's the only way there is. Seven hundred people from fifty
countries are huddled together in a barnlike auditorium on this collective farm between Minsk and Brest, 140 miles from the Polish border.
On the stage is a banner in English, which says:

WELCOME XIV INTERNATIONAL CONFERENCE
OF AGRICULTURAL ECONOMISTS

and another banner, much larger, which proclaims:

IN THE NAME OF LENIN
UNDER THE LEADERSHIP OF THE COMMUNIST PARTY
AHEAD TO VICTORY OF COMMUNISM!

Our Russian host introduces the chairman of this collective farm, a
rugged man probably in his forties, looking uncomfortable in his best
clothes. The chairman introduces his board members, who vary in
age and appearance. But all are convincing countrymen. One could
see how their fellow workers would have elected them. The names,
which I don't retain, seem mostly consonants.

The chairman speaks. There is translation, English only. Too bad
for those who speak neither Russian nor English.

"Kalinin farm is named after the first president of the Soviet
Union. The farm in its present form is twenty years old. Before the
revolution it was a large estate, the center of which was the village
which is now the headquarters of the farm. There are three thousand
people living on the farm, which has about ten thousand acres. There
are twelve hundred workers and thirty-eight administrators. The main

crops are wheat, rye, sugar beets, flax, potatoes, and cabbage. The leading livestock enterprise is dairy. There are also pigs, chickens, ducks, and geese. The farm is being converted to mechanical power, but there are still two hundred horses, one-third as many as twenty years ago. There are individual plots for families that want them, up to an acre in size. On its plot a family may have a cow, some pigs, and chickens. They may grow vegetables. This food can be consumed at home or may be sold in the market at the nearby city, at prices that are unregulated.

"Schooling is provided. Literacy is close to 100 percent. Health services are available, including hospital care. Retirement benefits are provided for the elderly. Social events are programmed for all, and athletics for the young. Are there any questions?"

"Is this an average farm?"

"We like to think it is better than average." And the visitors think so too, even without having yet seen it. We Americans know that when guests come to the United States, we show them farms that are better than average.

"What is your wheat yield?"

"About two metric tons per hectare" (29 bushels to the acre). This is less than in the United States but better than the average for the Soviet Union.

"Who decides what crops to grow and what livestock to keep?"

"The people, through the board members they elect." Some of the listeners exchange sidelong glances.

"Do the men work singly or in teams?"

"The people usually work in groups, as brigades, sectors, and teams. The women work as well as the men. Russian farmers have worked in groups for generations. On their private plots they work singly or as a family."

"How much money do the workers earn?"

There comes a torrent of Russian, which translates into rubles from various sources. Then there is produce for home use from the family plots. I give up and resolve to judge the level of living from what I can observe of housing, food, clothing, and health. Basic needs appear to be met, with the addition of a few amenities.

"Does everyone get the same wage?"

"No. A tractor driver gets more than a man who hoes sugar beets. The jobs are rated, and people get paid according to their work."

"Isn't that contrary to Marx's principle? He said, 'From each according to his ability, to each according to his need.'"

At this point the party man from the Ministry of Agriculture steps forward. He had been silent up to now.

"The principle we follow is, 'Better work and better results should receive a higher payment.'"

"How much of the Soviet Union's agricultural production comes from the private plots?"

"About one-third. These are not really private plots; they are part of the cooperative system. They make for good use of labor when the workers are not busy with farm operations. The fodder for the animals on these plots comes from the cooperative. The plots provide an incentive that increases the total production of the farm."

"Are people free to come and go? Can they leave the farm for a nonfarm job? Can they change from one farm to another?"

"There must be approval for such changes." The party man is frank. He knows that in his audience are scholars who have studied the Russian system.

"How do collective farms like this one differ from state farms?"

"Cooperative farms, which you sometimes call collective farms, are managed by a board of directors, as you have seen. State farms are managed centrally. State farms are closer to the communist ideal. Presently they produce a considerably larger share of our agricultural output than do the collective farms. We are increasing the role of state farms in our agriculture." (By 1980, two-thirds of the farm acreage in the USSR was in state farms.)

"Is food subsidized in the Soviet Union, or do consumers pay the full cost?"

To this question the party man gives a lengthy nonanswer. At least I am unable to discern an answer. I know from what I have read that there is a heavy subsidy.

"How are farm products priced?"

"The government is the principal wholesale purchaser. It announces the price in advance, and announces how much it will buy from each farm. The price for quantities in excess of the specified amount is 50 percent higher. So the board of directors knows ahead of time what the marketing prospects are."

"Does a farm worker receive a wage, or is his income based on the cooperative's profits?"

"Both. He receives a wage regardless of the profit or loss of the cooperative. If there is a profit, he receives a share of it also. Ordinarily about 70 to 80 percent of his income is from a monthly wage payment, and the balance, which fluctuates considerably from year to year, is from profit."

"Are you integrating your farms vertically?"

"We are. Increasingly we process farm products at the point of production. On this farm dairy products are processed. On other farms we process fruits and vegetables."

"Where does Kalinin Farm get its new technology?"

"We have research stations throughout the Soviet Union. Each farm has its contingent of agronomists, animal scientists, veterinarians, and other specialists. They obtain the new research findings from the research stations and help put them to use on the farms."

Some of the guests who understand neither Russian nor English are getting restless. The meeting is adjourned and we have lunch. This consists of bread, cheese, a hot stew, tomatoes, and cucumbers.

"Should we drink the water?" asks a first-time visitor. The water is safe. But our hosts have a solution for those who are apprehensive on this score: a bottled soft drink, name unknown to us, dark in color, fruitlike in taste, and omnipresent from the Polish border to central Asia.

After lunch we are dismissed to wander about and see the farm on our own. We note the terrain and the technology. The landscape is reminiscent of Wisconsin or Minnesota or even more northerly areas. This is not surprising; Kalinin Farm is of about the latitude of Winnipeg. Land is rolling. There are open fields, grazing lands, woods, lakes, and bogs. The soil is light gray, as is typical of northern soils formed under forest cover. It is not the highest in fertility. The amount of wheat straw, piled in the fields, gives evidence of a rather good crop. Sugar beets are doing fairly well. The cattle are Holsteins, well bred and in good flesh. The machinery is reminiscent of our John Deere and International Harvester lines, as well it might be: some of the Soviet designs are taken directly from the United States. Russian farm machinery tends to be larger and less sophisticated than ours—fewer things to go wrong. Farm buildings are adequate.

Apartment living is replacing individual units. New two-story apartment buildings are evident. To some of us, they rank higher from the standpoint of utility than they do with respect to aesthetics. There are still some log houses, with carved lintels over the doors and windows and with low, overhanging roofs. There are flowers in the dooryards of such homes, behind wooden picket fences. And there are embroidered white curtains at the windows.

On the hill overlooking the village is the falling mansion of the former owner, its vacant windows peering out through the tangle of what probably was once a lovely garden. Nearby stands a church, equally decrepit, its roof caved in. These two decaying structures are kept as reminders of the Communist victory over its enemies, the landlords and organized religion.

The village is a combination of the old and the new. Virtually everything here stands in service to Kalinin Farm. People move about, well dressed and apparently in good health. I expect that this farm often hosts visiting groups, and so the spectacle of seven hundred people from half a hundred countries is less of an event than might appear.

We note the traditional Russian vehicle: horse and wagon, the wagon with small front and large rear wheels, the horse in shafts joined by a yoke that arches high over the horse's withers. The horse is smaller than the draft horses of the West.

The weather is fine, with blue sky and white clouds. Everyone is enjoying it. Cameras click. We have been told we could take pictures of most scenes, but not of bridges, scenes of poverty, or private markets.

The men and women who do the farming are the most interesting to me. They are so authentic. Crops, animals, and soils seem built into their presence. Most of them are in the field, but some are about the village. The love of the Russian farmer for the land is traditional. Do they love it less because ownership is with the state rather than with themselves? Who can tell? Now and then there is a man who is without an arm or a leg. Did he lose it in the war or to a piece of farm machinery? These people are an updated version of the peasants Tolstoy wrote about. What is going on in their heads? I have a feeling that this is more important than all the statistics our hosts have quoted to us.

This is an area that has had a troubled history. Over the years the border between Poland and Russia has moved back and forth; the forefathers of these people have at one time been within Poland and at other times within Russia. For twenty-five years now they have had relative peace and a chance to build their society. There are few periods of like duration in Russian history.

I listen to the Russian language spoken in the street. I can't understand it, of course. But it is beautiful to hear. It has the harshness of a Russian winter and the softness of a Russian spring. The sounds are wondrously juxtaposed. There seems to be a love for the consonants. The words are clearly articulated, unlike the Romance languages of the Mediterranean. There is a cadence to the spoken word. It is not a courtly language, like the French; it is an earthy language. The sounds seem to come from deep in the throat, not high in the head. One can see how these people are poets; their language seems almost to compel it.

Not all of us are well clad for a farm visit. High-heeled shoes on some of the women, coats and ties on the gentlemen, not the best for seeing a dairy barn.

Many of the first-time visitors have been coached by friends on what to see. Some have been told they would see wasteful use of labor and leaky faucets and that they would witness lockstep mentality. These they search out and in due course find, to report back to their friends. Others have been told they would witness much social activity, equal sharing, and respect for law. These they look for and likewise find, and no doubt will likewise report. Almost any generalization about this enormous and diverse country has enough truth to make it credible and enough error to make it dangerous.

Really, this is not the best way to see a farm: a brief stop, a throng of people, and a language barrier. But our hosts are doing their best, and we see more than one might think.

It is time to go, and the passengers of our bus are finally collected. All but Mrs. Wykoff. Where is Mrs. Wykoff? Finally she comes, breathless and with eyes sparkling. She had made friends with a Russian woman and had been invited into a Russian home. And a pleasant home it was, as Mrs. Wykoff reports it. Electric lights, gas stove, gas water heater, refrigerator, rugs on the floor, television set, cut flowers on the mantle. And best of all a brief friendship, despite a total language barrier.

On our bus are mostly British, Canadians, and Americans. During the long ride back to Minsk we men talk over what we have seen.

"Not much freedom," says one. "Imagine having to get permission from the boss before you can change jobs."

"They've never had much freedom," says another. "They probably have more freedom now than they had under the landlords and the czars."

"It doesn't look much like socialism to me; it has a lot of capitalism in it," says another. "One-third of their production comes from the private sector, which is capitalistic no matter what they say. And they have what amounts to piecework, something our capitalistic employers have always wanted but our labor unions wouldn't stand for."

"They're changing," says a Britisher, a student of Russian agriculture. "Early on they had equal rewards, and they tried to abolish the family as a social unit. Give them time. We and they are on converging paths." This last observation is received in silence and with evident difference of opinion.

"There is a special thing about these private plots that the party man didn't mention, and you can understand why he did not. They provide a kind of pilot system for the planning process. When the officials develop their plan, they can take note of the free prices in these open markets and so get a clue as to whether to increase or decrease the production goal for a particular crop or livestock product. The socialist system needs a piece of the market economy to make it work."

"Just as the market economy needs a piece of socialism—public-supported research and education—to make *it* work." This sly comment goes without response, though it engenders considerable thought.

After a time someone says, "Kalinin Farm is run from the top, no matter what they say about democratic decision-making." No disagreement with that.

"One of the best things they do is to build nonfarm activities into their farming operations. Farm labor displaced by mechanization stays on the farm rather than going to Moscow or Minsk. They avoid the urban migration that plagues much of the world. They decentralize their industry, which is good from a defense standpoint."

"The farm is enormous. Does it need to be that big to be efficient?"

"No. They have gone beyond the point of efficiency. The reason it is so big is that it's easier to control one big farm than ten farms of moderate size."

"Their technology is behind that of the Western world, but they are advancing."

"They do better with crops than with livestock."

"I don't think they're doing so well technologically. In czarist times they exported wheat. Now they can't feed themselves. With the Lysenko affair they held their life sciences back for a whole generation."

(Trofim D. Lysenko was a plant breeder who, with support from Stalin, forced Russian scientists to follow a discredited theory of genetics—the inheritance of acquired characteristics. He exercised great power from about 1935 to 1964. Thirty years of nonscience left the Soviet Union with an outmoded agriculture.)

"They seem to have a functional system, flawed though it is. But think of the millions of people they had to starve or kill or exile in order to destroy the old system and make room for the new."

"True, but that's not the question. That's all history. These people had no part in that business. All they're trying to do is to improve what they inherited, which is all we're trying to do."

"But it *is* the question for any country considering the adoption of the Communist system."

And with that we all agree.

"The level of living on Russian farms has come up," says the British scholar. "In the early years the commissars held farm incomes at a low level, put little capital into agriculture, and fed their people on bread and potatoes. They extracted forced savings out of agriculture to invest in industry, science, and a military buildup. In recent years they have gone more toward a consumer society, with better diets and higher farm income. You can see evidence of that at Kalinin Farm. Levels of living are now much the same, farm and nonfarm."

"I don't believe what we saw there. They showed us their best."

"Some people predict that the socialized agricultural system of the Soviet Union will collapse, either because of limited freedom or top-heavy bureaucracy or lack of incentives. What do we think after seeing Kalinin Farm?"

No one thinks the system is due for an early collapse.

"Why do they move toward state farms if the collectives are doing well?"

"Probably because state farms are directed by one man and so are easier to control."

"The enterprise system makes many small changes and so stays on course. The centrally directed system is slow to change, and when it does, the change is abrupt. Both its successes and its failures are likely to be spectacular."

"I like the people, but I dislike the system," says a man who has been silent up to now. There is no disagreement. This is probably the most frequent comment made by Western visitors to the Soviet Union.

The conversation wanes. We have exhausted either the subject or ourselves.

At the front of the bus, Natasha, the Intourist guide, is answering questions. She is the only Russian we really get acquainted with on this trip. She is probably in her twenties, tall and dark. She flutters her long eyelashes while she thinks up an answer to a question. She knows nothing of agriculture. Today she guides our group. Next time she may guide a group of engineers. On yet another day she may guide a busload of retired folks from Paducah. She is well programmed.

"Our ideas are different from those of our parents," she says, "but their ideas are very dear to us."

"I passed my examination on the United States this spring. We had to know every state and capital."

"What is the capital of Ohio?" one of us asks.

Her eyelids flutter. "Oh! I don't remember. I have now forgotten everything."

"I thought the delegates would be old men with beards, instead of so many young men."

The men, young and old, seem more attentive than do the women.

"Why do so few American students study Russia?" she asks. "I think it is nice to study."

Someone inquires, "Who is your favorite American author?"

"Mark Twain," she says. "I love Tom Sawyer. He is the favorite hero of the Russians. I like Jack London, and I like Robert Burns." She quotes from one of Burns's poems.

So we make our way back to Minsk, a fleet of a dozen buses, complete with escort on the road and in the air. We arrive at our hotel tired, hungry, and somewhat confused. Anyway, we saw a real Russian farm.

Or did we?

The United States

Family Farmer

Delphi, Indiana, November, 1982. For the magnificence of its agricultural endowment, nothing in the world matches the American Midwest. Stretching from the Appalachians to the Rockies and from the Gulf of Mexico up into Canada, a thousand miles or more in each direction, this is the largest of the world's garden spots.

Consider how good this area is. The topography, long ago sculpted by the glacier, is on the whole adapted to modern machine tillage. Soils are deep and rich. Rainfall is generally adequate except in the West. The winters are severe enough to reduce the disease and insect problems found in the tropics. Summers are long enough to mature a good crop. Population per square mile permits farms sufficiently large to be efficient.

The area is penetrated by great waterways, providing the best of transportation for farm products. Oceangoing vessels load wheat at Duluth, fifteen hundred miles from the sea. The Mississippi system has more than six thousand miles of river transportation reaching into the heart of the Midwest; its fingers extend up the Ohio, the Missouri, the Arkansas, the Tennessee, and the Illinois all the way to Lake Michigan. Corn, soybeans, wheat, and other farm products go down the Mississippi, out the Great Lakes, and overland to the Atlantic and the Gulf Coast to help meet the food needs of most of the world's 160 countries.

The people who penetrated the American midland had the vision to foresee its possibilities. They insisted that it be settled in a pattern of private individual ownership rather than in the tenancy patterns of Europe, which they had left behind. They provided for education, they began experiment stations, they set up the extension service, they put together drainage districts and improved waterways. They criss-crossed the area with roads and railroads. They fought for and got good mail service, rural electrification, and rural telephones. They set up grain elevators, markets, warehouses, stockyards, and processing plants. A whole new set of agricultural industries was created: farm machinery, fertilizer, motor fuel, and agricultural chemicals. They set up credit institutions to tap the eastern money markets and so bring investment capital into the new area.

In managerial skills the people who farm this area are among the most capable in the world. Good basic education, good agricultural research, good adult education, and the incentives latent in entrepreneurship—all of these they have.

At harvest time this midwestern cornucopia pours out a flood of corn and soybeans, especially so in a good year. This year, 1982, produces the best crop the Midwest has ever had. It is the year the pictures in the seed catalog come true.

On the drive out to the Mills farm the November sky is gray, the air cool, and the breeze mild. Soybeans have been combined. Three-fourths of the corn has been harvested, the leveled and chewed-up stalks contrasting sharply with the straight rows still standing. Bins bulge with the harvest. Cattle graze on hillsides green from recent rains.

The highway dips down to cross the Tippecanoe River a few miles above where William Henry Harrison fought the Indians in 1811. The road rises and then dips again to cross the Wabash, a bigger river. Sycamores grow along the bank. Oak, maple, hickory, and tulip trees, now mostly bare, cover the slope from the valley floor to the upland. North of Delphi the road recrosses the Wabash, climbs the slope, and levels off. A mile further on is the farm of Bob and Dorothy Mills.

The Mills farm consists of nine hundred acres of corn and soybeans on the upland and one hundred acres of wood along the river. This is a three-family operation: Bob Mills, fifty-two years old, and his wife Dorothy; son Mark, twenty-two, and his wife Michelle, married just this past summer; and son-in-law Ed Oilar, twenty-eight, and his wife Marsha. Both Ed and Mark live within a mile of Bob and Dorothy. There are two other daughters, both married to nonfarmers and both working. Debbie is a music teacher at nearby Twin Lakes, and Lois is a primary teacher in Champaign, Illinois.

Bob's mother, a widow, lives one hundred yards up the road, on the old home place. Her husband died recently; she appreciates being near her children, her grandchildren, and her great-grandchildren.

The land lies well, gently rolling. It was plowed by the glacier twenty thousand years ago; that uninvited guest left a few calling cards in the form of boulders, some as big as a barrel. Years ago this area was a transition zone which lay between the prairie, stretching west to the Rocky Mountains, and the woodlands, reaching east to the Atlantic Ocean. Now the black prairie land is almost all in crop. The forest has been felled except for the river breaks, the steep slopes, and an occasional low spot. The forest soils are light in color and somewhat easier to work. Both prairie and forest land are deep and fertile, though they must be handled differently.

Bob is a pleasant man, frank in his conversation and friendly with his guests. Dorothy is brisk and genial. Her special love is music.

"Has the farm been in the family for a long time?"

Bob replies, "My great-great grandfather, John Pearson, bought all the land in this area about 1835 or 1840. He split it among his four children. My branch of the family got this tract."

Dorothy says, "My great-great-great grandfather was the first settler in what is now the city of Delphi, and one of the first two settlers in Carroll County. He came in 1826." That was fifteen years after the Battle of Tippecanoe.

"Did you always know you would farm?"

"No," says Bob. "I decided not to farm; I'd had enough of it, growing up on this very place. I decided to be a jeweler and went to California to learn watch-repairing. I returned and worked in a jewelry shop in West Lafayette, where Dorothy went to college. We had been high school sweethearts. We got married in 1950. I got into the Army and went to Germany. In 1953, when I came back, my father had some health problems and wanted to retire. He asked me if I would take over the farm. We began farming my father's 147 acres, on a rental basis."

"I said I would never marry a farmer," Dorothy says, "but I did." She was graduated from Purdue and taught home economics for one year at Carrollton High School.

"What was it like, getting started?"

"It was hard going. I began with corn and hogs, both of which I knew from my upbringing. We both worked hard and saved what we could."

Dorothy remembers one time when she was driving the tractor, cultivating corn in a field with point rows. She came to the edge of the field, bordering on a ditch, and didn't know how to make the turn. Bob found her there in tears. "I quit!" she said. "We hired Ed Oilar sometime after that," says Bob. Dorothy continued to keep the books, a job that grew as time went on.

The farm wasn't big enough, especially on a rental basis, sharing returns with the landlord. So in 1959 Bob rented another 90 acres. In 1963 he added 100 more acres, also rented. In 1970 he bought his first land, 120 acres. In 1977 he bought 134 acres, and in 1978 he bought 80 acres more. In 1981 he rented and operated 300 additional acres owned by Dorothy's sister. In 1982 he is farming altogether about 1,000 acres, one-third of it owned and the rest rented. Most of the land lies close to the homestead. The farm of Dorothy's sister is three miles off.

"How many farms were there at an earlier time, on the land you now farm as a unit?"

Bob reflects, counting on his fingers. "Seven," he says.

"What became of the six farmers that were displaced?"

"Several of them retired. The next generation didn't want to farm, so they sold their farms to me. A couple of owners kept the home place and still live on it but rent me the fields. Some of these people now work in town."

"Are they better or worse off for having left the farm?"

"Everybody made his own choice, so at least they thought they were better off."

"How did Ed Oilar and Mark get started with you?"

"As the farm got bigger, I needed a hired man. Marsha and Ed had been dating. Marsha said Ed would be a good worker, and her mother agreed. So I hired him. They were right. Ed had been to Indiana State for his education. He had been assistant manager for a lumber company and had no farm background. But he learned fast. In 1977 he and Marsha were married. They live in a mobile home up the road a bit. Ed is excellent with livestock. He took a winter short course in agriculture at Purdue. He has a share in the hog enterprise and is working into the business."

"And Mark?"

"Mark always wanted to be a farmer. He also took a winter short course; maybe he'll take another. He is good with the crops and the machinery. Like Ed, he lives in a mobile home. It's a pretty place, in an oak grove with the pond close by. We'll see it after lunch. Mark has an enterprise of his own. He's district dealer for Jacques Seed Company. He's working into bigger responsibility. Both Mark and Ed are starting small like I did, and are growing into it. But they start with a higher level of living than Dorothy and I had."

"How hard is it for a young man to get started farming these days?"

"Pretty hard unless he has family help. He could rent land if it is available, and he could buy used machinery. Even so, it would take close to one hundred thousand dollars to buy a line of equipment. Some of that he could borrow, but with interest rates and grain prices what they are, he might have a real cash flow problem."

How things have changed! I remember my father telling about how he started farming in northwest Iowa in 1908, on a patch of rented land, before he was married. He had only a team of horses, a wagon, and a few tillage tools. He needed enough ready money to live on until he made a crop. His capital outlay was measured in a few hundred dollars, not in many thousands. Each of his six brothers got started farming on about the same scale and with similar resources. All became established owner-operator farmers.

I wrenched myself back to the present. "Would you describe the hog operation?"

Bob is proud of his hogs. "We sell about eleven hundred head a year, seven hundred of them farrow-to-finish, and four hundred as feeder pigs. The sows are my own stock. Males are Boar Power. I don't have fancy equipment. Sows are bred in outside lots. We farrow in a remodeled barn and a converted chicken house, four farrowings a year. We wean an average of about nine pigs per litter: good care cuts down on the mortality. A veterinarian runs a regular check to keep the hogs healthy. We sell to Wilson at Logansport on a grade-and-yield basis."

We later see the hogs. They are smooth and thrifty, their different colors reflecting the varied bloodlines that contribute to good rates of growth.

"Do you have cattle?"

"We used to. But we gave it up. Hogs make more money."

"How about corn and soybeans?"

"We grow about 450 acres of each. We have grown some wheat, but not this year. We fall plow the heavier ground for corn, turning under about 500 pounds of fertilizer per acre. We put on about 150 pounds of nitrogen in the spring and use 200 pounds of starter. I apply my own insecticides and herbicides; I think I can do a better job than the custom operators. We feed about half the crop and sell the rest."

"And your yield?"

"This is our best year. My different fields will run from 140 to 195 bushels to the acre. Average for the 450 acres is probably 160 bushels."

Fifty years ago, when I was a young man, the Purdue Extension Service had a Hundred Bushel Corn Club. A farmer would select his best five acres, apply the 125 pounds per acre of 2-12-6 fertilizer then recommended by the county agent, manure the land heavily, and follow the best tillage practices of the day. If he grew one hundred bushels to the acre, he got a gold medal. Few there were who made it. I tried it, and got only seventy bushels. If we had a Hundred Bushel Corn Club today it would take much of the gold in Fort Knox to provide the medals.

Later I rechecked the grain yields on some of the other farms we have seen. Of course, these are individual farms, not national averages. The comparison is shown.

	Yield per acre per year, metric tons
Bali, Umedesa Subak, 2 crops rice, 1 crop corn (irrigated)	7.0
China, Yuan Jia Brigade, 1 crop rice, 1 crop corn (irrigated)	5.0
U.S., Mills Farm, corn (rainfed)	4.0
Taiwan, Tsai Farm, 2 crops rice (irrigated)	2.9
India, Chandra Farm, wheat (irrigated)	1.2
El Salvador, Salinas Farm, corn and beans (rainfed)	1.1
USSR, Kalinin Farm, wheat (rainfed)	0.8
Portugal, Silveira Farm, oats (rainfed)	0.4
Upper Volta, Boureima Farm, sorghum (rainfed)	0.3

Bob isn't in the government corn program this year. It involves restricting acreage in order to qualify for payments, price supports, and help with storage. He expects to be in the program next year. He will be in or out of the government programs depending on whether it appears to be to his advantage. He is neither for nor against them as a matter of principle.

Bob's soybeans average a little over fifty bushels to the acre, well above the United States average of thirty-five bushels. Some are drilled solid, some in rows.

Tillage is mostly conventional. The land is fairly level, not erosive. Some fields are chiseled. The moldboard plow is used for the heavy ground.

Bob has three big tractors and two smaller ones, John Deere and White. All of them have seen considerable use. He has no four-wheel-drive tractor. He has a big Allis-Chalmers Gleaner to harvest corn and beans. He has both bin and batch grain driers.

The farm is incorporated. It is a family corporation, like any other family farm except in legal form. Bob and Dorothy own a majority of the stock; the children own the balance. Each has the same share, whether son or daughter and whether or not involved in the farm operation. "I've had experience with an estate that was not planned," Bob says. "We have worked it out so that the problem of transferring title to the next generation should be minimal."

It is an authentic family operation, three couples and two generations. The generally accepted definition of a family farm is that it is a farm on which the majority of the labor and management is supplied by the farmer and his family. On that basis the Mills farm clearly qualifies.

Each individual has something of his or her own. Bob is president of the Rural Electric Membership Cooperative and has served on several boards. Dorothy gives piano lessons three afternoons a week, has been president of the Parent Teachers Association and was active with Band Boosters. Ed has, on his own, a seed corn dealership and a piece of the hog enterprise. Mark has his seed business. Marsha, Ed's wife, is a registered nurse at White County Hospital. Mark's wife, Michelle, is a secretary at an accounting firm.

"How are you doing in this year of low grain prices?"

"Not so good. But we'll make it. I've been careful not to get overextended. We've grown slowly, by steps. We haven't plunged or leveraged ourselves like some people have. I have my debt load at a level that can be serviced, even if times are on the hard side. We're diversified; about half our income is from hogs, and half is from grain. Often if one is down the other is up. This year grain is cheap, but hogs do fairly well. I rent two-thirds of the land. There is more flexibility in a rental arrangement than in a mortgage with fixed debt service. And all of us have some off-farm income."

The Mills thoroughly enjoy themselves. Bob is fairly relaxed; he regularly takes a nap after mid-day meal. Sunday is for church and for rest, not a workday except possibly at planting or harvest time. Bob and Dorothy have taken a vacation every year since they were married, even when they had to borrow money to do it. One year they went to the Canadian Rockies. In the winter they go where it is warm; this year it probably will be Mexico. Mark and Ed take full responsibility while Bob and Dorothy are away. In the woodlot near the homestead is a cabin; families from the neighborhood gather there each Fourth of July for a picnic. The pond at Mark's place, spring-fed and stocked with bluegills, bass, and catfish, provides fishing and skating.

"How do the three families get along?" I ask the straightforward question.

"We get along well; we make it a point to do so. We've always encouraged the children to be frank, even when they were little, and they are, including the in-laws. My dad allowed me to make mistakes so I could learn from them, and I do the same with Mark and Ed. Each family has its own house, near enough together to work together in the farm business but far enough apart so as to live their own lives. Two-generation farming isn't for everybody. Where it works, it's wonderful; but if there is going to be constant friction, it's best not to try it."

"Do you worry about big corporate farms' taking over from the family farm?"

"No. Farms much bigger than ours run into all kinds of problems. Often they have an enormous overhead for labor and interest. In a year of low prices like this one, they're in trouble. With paid management and labor there isn't as much incentive as there is for us. We can hold our own."

"Will you get bigger as Ed and Mark grow into more responsibility?"

"Probably some. But I'd like to see the farm stay of a size that we can handle ourselves. That way we can do the job the way it should be done."

Thereupon Dorothy calls for the mid-day meal. Ed has shot a rabbit. The main dish is rabbit stew, cooked with onion, apple, and mushrooms. It is delicious. We pitch in for what amounts to a pre-Thanksgiving dinner.

Upper Volta

Two Hats

Nedogo, July 21, 1982. This is the Sahel, Africa's trouble zone. To the north are the dry sandy wastes of the Sahara. To the south is the well-watered lush green of the tropics. Between, running irregularly east and west from the Atlantic to the Nile and beyond is a transition belt about five hundred miles wide, roughly between ten and twenty degrees north latitude. Annual rainfall at the north border of this zone is about ten inches a year; at the southern limit it is about thirty-five inches. In a hot climate with much evaporation, this is dry country.

In the early days this was savanna with brush and sparse vegetative cover. Grass was thin in the north, more productive in the south. Rainfall was erratic as well as insufficient, then as now.

The native plants were adapted to this setting. The baobab tree stored its own water; the leaves of the bush plants were waxy, resisting transpiration; the grass could shrivel but stay alive; seeds could lie dormant in the soil until the rains came. The native animals were also adapted. If forage was scarce, the antelope could move out.

These patterns of climate, plant, and animal life are found in various places in the world besides the Sahel: in the high plains of Texas, the steppe of central Asia, and on the borders of the Kalihari Desert in southern Africa.

When men first came to the Sahel untold centuries ago, they adapted themselves to this setting. They were nomadic herdsmen. If rains were plentiful and forage abundant, they increased their herds. In time of drought they moved their herds, or, if the drought was general, they killed the animals and ate the meat.

But a grazing system doesn't produce much human food and doesn't support many people. Tilled crops can increase the production of human food by several orders of magnitude. As population increased, people were forced into substituting a cropping system for grazing. The people became sedentary rather than nomadic. Food production was increased, but the flexibility of the system was reduced. In time of drought the crops failed. With a system based on crops for human consumption, there were fewer animals to be liquidated. The food supply had lost its cushion, its shock absorber, its adjustment factor. It was hard for a settled people to pick up their village and leave. Where would they go? So came the classic problem of the transition zone: stable food needs, variable food production, and an agricultural system that had lost its capability to adjust. Marginal resources were stretched toward their limit.

Upper Volta, a nation of the Sahel, is one of the poorest countries in the world.

Ouedraogo Boureima is a citizen of Upper Volta, that landlocked country in West Africa, about the size of Colorado. Boureima is trying to cope in a difficult setting. He is of the Mossi tribe, as indicated by the deliberately inflicted scars on his forehead and cheek. He is a farmer, fifty-one years old, very black and very intelligent. He never attended school a day in his life. He is six feet tall, lean and tough. He has a sparse whitish beard. He wears a green and brown knitted cap. He has green baggy trousers, a long light tan coat or robe, and sandals with leather thongs. This man brings chairs for his guests, chairs made in the village from saplings woven and dried. He seats himself on a log under a spreading tree. The soft African breeze moves through. Nearby are round adobe huts with conical thatched roofs, one hut for each of his four wives. Unthreshed grain is stored in large woven basketlike containers, held off the ground with wooden supports. Threshed grain is saved in huge red clay pots. Grain is ground by pounding in a hollow tree trunk with a heavy wooden mallet or by crushing with a roller in a hollowed stone.

This is a village of about two hundred families, clustered together yet not seeming overcrowded. Goats, turkeys, chickens, and guinea fowl wander about as we talk. Villagers pause to listen a while and then pass on. Beyond the village are fields of corn, sorghum, and millet. The crop was planted in May and June, when the seasonal rains began. It is now July, and the crops are well up, off to a good start. Women and children—and some men—are weeding the fields, using the short-handled hoe called a *daba*. The soil is yellow, brown, and red, sandy and thin, with occasional rocks. The landscape is gently undulating. Cropping is done under and around trees and shrubs.

Because three-fourths of the people in the village have the same surname as our host, Ouedraogo, we address him by his given name. He is Mr. Boureima. He speaks his native Mossi tongue, Morre. A fellow Mossi, Seydou, who has been to school, translates from Morre to French. Joan Brakke, a Peace Corps volunteer, translates from French to English. Mahlon Lang, a Purdue University economist, puts the questions.

"Tell us about your farming operation."

As nearly as we can figure, the total area comes to about twenty acres. The Mossi people have no standard land measurement. A few acres are in "house field," near the village, relatively fertile with night soil and animal manure. Most of Boureima's land is in "bush field," four miles distant. In addition there is a "bas field," lowland along a creek that is often dry. Boureima's farm is much larger than average. He has two oxen, one or two burros, some goats and chickens. He owns a piece of animal-drawn equipment that looks like the shovel-plow I followed while a farm youth in the American Midwest sixty years ago. He has a labor force of eight people: four wives, three children of working age, and himself. Labor is rarely hired or hired out.

Boureima grows corn (maize) in his house field. He grows white and red sorghum and millet in his bush fields. He has some peanuts and some *roselle*, a legume. All of these crops are for human food in some form or another. The animals get a small amount of grain when they are working. Otherwise they subsist mostly on green forage, waste, and crop residue.

"How did you get started farming?"

"I farmed with my father. I became responsible for the farm when I was thirty years old. I got married when I was thirty-five. My father died five years after that. Then I farmed for myself."

"What did you learn from your father?"

"I learned to know, by the spring moon, when the rains would come, and so when to plant."

"Did you know you were going to be a farmer?"

"Yes, my father taught me it was good to be a farmer. He said, 'If you don't learn to become a farmer, you will become a thief. Farmers live well; thieves do not!'"

"What do you do that is different from what your father did?"

There comes a long explanation of how he had made the change from exclusively hand labor to animal traction, a matter in which he takes much pride. This change should not be underestimated; our European ancestors took many generations to make it. Some French people from the Organization of Regional Development persuaded him to try a piece of animal-drawn cultivating equipment. They showed him how to harness the burro and guide the cultivator. There was a small charge which Boureima paid. Four others tried the tool, but only Boureima persisted. With this tool he and his wife could feed ten people. Seeing his success, others became interested. Now animal traction has become somewhat more widely used in the village, though it is still a rarity in the country. One problem is feed for the animals, not a small thing where human food is in short supply.

Boureima is making changes. He is experimenting with "tied ridges" which make shallow, rain-retaining basins between the crop rows instead of the long furrows that allow the water to run off. This is with the help of the United States Agency for International Development.

These villagers are resourceful. For example, there is use of the bas field, a flood plain along a creek. In this field they plant both rice and sorghum. If it is wet, the water rises, the sorghum is drowned, and the rice flourishes. If it is dry, the rice dies out, and the sorghum makes a crop.

"Tell us how you meet your need for water."

"There is a pond in the bush. A guard watches to see that everyone who takes a pail of water first removes a pail of mud to make the pond deeper. One month of the year the pond is dry. Then we carry water from a well two miles away. If a bad drought comes, both pond and well fail. Then we dig a deep hole in the dry stream bed and wait three or four hours for enough water to seep in so we can dip it out."

"How did you manage during the bad drought some years ago?"

"The crops failed. I had grain stored for my family. Those who had stored no grain sold animals and bought grain. Some had relatives who could help. Ten people kept alive on the rations of two. Many people died."

We learned from other sources that Upper Volta has almost no food relief program of its own. The Catholic Welfare Service does an ongoing job, expanding during emergencies. During the great drought of 1972–75, large amounts of relief food came in from the United States under Public Law 480. Multilateral food aid was provided by the World Food Program of the United Nations. The bad years seem to be clustered, in unpredictable fashion. The good years also appear to cluster. The time of our visit is during one of the better periods.

"How do you guard against another bad drought?"

"Plant more crops. If some fields fail, others may yield enough so we will have food. Store grain. Keep animals to sell and buy grain."

"What kind of food does a person need to be healthy?"

"Best food is tow (a thick sort of gruel made from maize, sorghum, or millet). Sauce is good if you can get it. (Sauce can be meat with a condiment, grain of roselle, a legume.) Peanut is good. And fruit."

We learn from the rural development people that 90 or 95 percent of all the food that is produced is consumed in the village. Only a very small amount is sold.

"If there is a dispute about land, how is it settled?"

"There is a 'chef de la ordu,' the oldest man in the village. He knows the history of each field and each family. If there is a dispute, he listens and talks with his counselors. Then he decides, and it is settled."

Boureima goes on to tell us the organization of the village. A nuclear family forms a "menage." A number of these make up a "concession." Concessions combine into a "cartier." Together these comprise the village, which has a chief. The law requires that the chief be elected, but the people always elect the son of the former chief. The chiefs from a particular area have a "zonal chief," and the zonal chiefs have over them a "paramount chief." All of this is by tribe. The Mossi tribe is the largest in the country.

We learn that the spiritual lives of the people find expression in a number of religious forms. Moslems are the most numerous. In addition there are Catholics, Protestants, and animists as well as a combination of these. Boureima is a Moslem. This is the time of Ramadan; neither food nor water is taken between sunrise and sunset.

Boureima's first wife is forty-one years old. She has given him three children of working age. Recently, having prospered, he has taken three additional wives, the maximum allowed by Islamic law. Each of the new wives is in her twenties, and each gave birth to a son this past year.

With some men taking multiple wives, there is a shortage of women available for marriage and so a substantial bride price. Men generally wait until they are thirty or thirty-five before they marry; it takes time to accumulate the bride price. Women are usually married early, some at age fifteen. The government has a family planning program which has some acceptance in the city, but is virtually without effect in the country, where children are considered an asset. About 87 percent of the people in Upper Volta are farmers. A country woman may bear six or eight or ten children. If a wife does not quickly produce children, the man may take another wife.

We learn that medical services are increasing in availability. Formerly when a villager was sick, he sought out the herb doctor. Now he can go to the dispensary at Pabre, six miles away. With this improvement in medical service, the death rate has fallen. But the birth rate remains high. The annual rate of population growth is 1.8 percent. If that rate continues, in forty years the population will double. These facts and their significance are not clearly perceived by the village people.

When a person is too old to work, our host tells us, a child from the family is given the task of attending him.

The school nearest Nedogo is at Pabre, six miles distant. From the village of Nedogo with two hundred families, only one child attends school. Instruction is in French. The Morri children are bewildered by the strange language; few who begin persist beyond a year.

"If there were a school here in Nedogo, would more children go?"

"Yes! And I would go too!"

Recently "farmer schools" have been started by the government. The schoolmaster receives a small salary and the use of a tract of land. The scholars work on the land, learning farm skills. In addition they learn enough French so they can communicate a bit. And they learn their numbers, at least a little. The schoolmaster supplements his meager income with earnings from the farm. We hear that these schools are more a bridge for getting out of agriculture than a preparation for staying in.

We go out to the bush field. Here millet and sorghum are growing. Yields per acre here are relatively low, maybe one-third as much as in the high plains of Texas. Two of Boureima's sons are cultivating sorghum, one leading the bullock while the other guides the cultivator. The four wives are working together, weeding sorghum, bent over, using that all-purpose tool, the daba. The children play under the shade tree, the older tending the younger. The wives leave their work to nurse the babies.

The soil in this distant field is thin and poor, low in nitrogen and phosphate. Everything is taken off, grain, stalks, and all. The field is too far from the village to receive night soil or animal manure. In former times rotating tillage was practiced; a field would be farmed for some years and then would go back to bush, to rest, accumulate organic matter, and regenerate its productive strength. But now the population has grown so that most of the tillable land in the area must be used each year.

The land here was cleared from the bush in fairly recent years. Older people tell of lions and warthogs. We attend, in Ouagadougou, a native dance in which the animal tradition is strongly in evidence. But wild game has disappeared.

We see the fertilizer trials. A unique problem arose in this experiment: with the operator illiterate, how should the fields be designated and marked? An inspiration: identify the plots by color—red, blue, yellow, orange, and so on. But here a difficulty presented itself. The Mossi people have perceptions and names for colors, but these are few and apply to bands of the spectrum that may differ from ours. There is green (crops), blue (sky), and yellow (soil). But they have no names for or interest in those colors that are to them unimportant.

It is too early to see marked differences in the fertilizer trials. We see a patch in the field, maybe twenty feet in diameter, where the sorghum is deep green and twice as high as in the surrounding area.

"It was an anthill, now knocked over and leveled," we are told. The ants dug deep into the earth to bring up moist subsoil with which to build their mound. In so doing they opened up passages through the hard earth. They brought in organic matter for themselves and their young. What are the ants trying to tell us? That the surface soil is exhausted? That the hardpan should be broken up so the roots can penetrate? That more organic matter is needed? All of these things? A biblical passage comes to mind: "Go to the ant, thou sluggard. Consider her ways and be wise" (Prov. 6:6).

We see another patch of flourishing growth, about the size of a tennis court. "Probably was a Fulani corral," we are told. The Fulani were nomadic herdsmen who ranged over this land before the Mossi settled it. They had corrals and stockades where the animals dropped manure.

We ask Boureima what might be his hopes for the future. What would he like to see happen if he could have his way?

"That people would stop drinking beer," he says. "It is a waste of money and it makes people crazy." If we Westerners had been asked to draw up a list of hoped-for happenings, this probably would have been well down toward the bottom. Some of the difficulties of agricultural development arise from trying to solve what are not perceived as problems by the intended beneficiaries.

"Will your sons be farmers?"

Boureima's face clouds. 'They must farm. They have no schooling. But there will not be land enough for them in this village. They must find land somewhere else."

"What is the greatest problem you see?"

"Lack of knowledge to use better methods. Getting materials, like fertilizer. Getting credit."

There are problems that our host does not name, maybe does not see: excessive population growth, loss of soil fertility, depleted firewood. In moments of candor the rural development people confess that they feel as if they were trying to sweep back the ocean.

There are observers who combine these problems into a worst-case scenario. But the worst-case scenario is not the most probable event. Great difficulties have been overcome before. Maybe good things can be made to happen. Maybe Boureima's tied ridges will conserve moisture. Possibly, drought-resistant sorghum will be bred. Malaria, bilharziasis, and river-blindness are being overcome, so that well-watered lands in the south can be opened for farming. (Bilharziasis, also known as schistosomiasis, is a debilitating parasitic disease carried by snails found in contaminated water. River-blindness has as its vector an insect that finds its habitat along watercourses. Malaria is mosquito-borne, as we all know. All of these diseases have water association and are found in tropical Africa.) Surely, in time, family planning will catch on.

The people of Nedogo are front-line famine-fighters. They have shown resilience and resourcefulness. They do not seem to feel defeated. Survival has always been the test, and a greater proportion of them are surviving than in time past. Infant mortality is declining, and the life span has lengthened.

The satisfaction index is measured by achievement relative to expectations. The proper strategy in economic development is to lift expectations enough to stimulate enterprise, but not so far beyond likely accomplishment as to produce despair. To lift achievement enough to make enterprise visibly rewarding. To gauge these things as do the people concerned, not to impute to the society the values of one's own world. And, ideally, to do all this in a way and at a pace that does not tear the society apart. If there is a more challenging job, I don't know what it is.

There has been little change here for thousands of years. We later see, on the monuments of Egypt, carvings of farmers plowing with oxen twenty-five hundred years ago. The Mossi tribesmen are just now beginning to use animal traction. Why did one culture advance while the other remained unchanged? This is the riddle that agricultural development workers find difficult to answer.

As we leave, we exchange gifts. Boureima presents the traditional Mossi gift, a live chicken. We give him a nontraditional gift, a gold-colored cap with the emblem of Purdue University on its peak. He removes his brown and green knitted cap and puts on the new one.

There is an appropriateness in his having two hats, one a symbol of tradition and the other signifying the world of change. This man is a bridge between the old and the new. He is what the agricultural development people call a "change agent." The changes he is making break customs that are centuries old. The developing world needs two-hat people. We hope he keeps both hats.

Epilogue

As stated at the outset, interviews with these farmers are case studies, not suited for generalization. For each of the farmers interviewed, there are in the world some millions who went unquestioned. Nevertheless, one cannot visit farmers in so many countries without receiving some general impressions.

Throughout our visits there was an unintended search for a norm. Questions that continually thrust themselves forward were these: Is there a unique worthiness in agrarian life? Should farms be large or small? Should they be single proprietorships, cooperatives, corporations, or state farms? Should land, management, and labor all be in the same hands, or should they be separately supplied? Who should make the decisions about crop and livestock production? Above all, by what criteria should these things be judged?

As regards farming systems, there came these unbidden but indelible impressions:

Size. Agriculture involves people, plants, and animals. These living beings accommodate themselves poorly to the inflexible and impersonal setting that seems to be associated with a huge farm. By breaking large farms into smaller decision-making units, as the Chinese now do, the handicaps of size may be reduced.

At the other extreme, the disadvantages of tiny farms are quite obvious, and, with the increase of population numbers, a greater problem.

Incentives. If a farmer is to perform well, he must see a link between his effort and his reward.

Credibility. Farmers must believe in the system if it is to work. Such confidence is a necessary but not a sufficient condition for success.

Efficiency. This is important. But it cannot be the sole objective of a farmer or a farming system. Society insists on broader criteria.

Diversity. There is no need for a country's farming system to be monolithic. No single system can satisfy the diversity of conditions in any one country.

The true test of a farming system is not its conformity to some ideology. A better measure is its achievement judged against a multiple standard: economic performance, ecological soundness, and political acceptability. If it fails by any of these criteria, it is due for change.

Change is occurring. In fact, change is the norm. The Chinese agricultural system experienced an enormous change. El Salvador is undergoing a recent and lesser change. The systems in Portugal and Upper Volta are under stress, the one political and the other ecological. The Taiwan system, which is resilient, appears to be coping.

Farmers are not the only famine-fighters. There are also the agricultural scientists, the institution builders, the family planners, and those who work at famine relief. Together they may overcome world hunger. But those are other stories, subjects for other books.

Acknowledgments

Major thanks go to my wife, Eva. She and I visited most of these farms together. She made notes, took pictures, kept files, worked out logistics, asked some of the better questions, and spread good will, of which she has an abundant supply. She consulted on the form of the writing, edited the various drafts, and did repeated typing.

A number of our good friends read and commented helpfully on an early draft of the manuscript: Lowell and Mary Hardin, Fred and Ann Warren, and Betsy Schuhmann.

Other credits are as follows, by country:

Bali. Raka Djaja, Ineugah Gedoa, A. Tjatern.

Brazil. Eilseu Alves, Paulo U. V. Caldeira, Ellsworth Christmas, James Cullom, Homer Erickson, João B. Ferrer, Olivier Silva da Magales, Murdock Montgomery, Victor Jose Pellegrini.

China. Bian Jiang, C. C. Chang, Madame Chen Cui Xian, Fan Tiah Min, Madame Feng, Gu Zhong Nang, Hao Xiang Qian, Ji Hou Fu, Madame Jin Yun Xiu, Liu Cong Meng, Madame Liu Gan Li, Liu Wei Zheng, Liu Yu Liang, Richard Lyng, Ma Hui Lian, Leo Mayer, Shen Hao, Shen Jin Pu, Richard Smith, Madame Wang Shun Ying, Wu Shao He, Xian Zhu Fang, Xie Yong Sheng, Yang Lian Fang, Ye Ping Wen, Mr. Zhang, Zhang Hong, Zhang Jian, Zhang Ji Wei.

El Salvador. Peter M. Cody, Eduardo Hipsley, Ronald J. Ivey, Tim O'Hair, Roberto Rosales, Gail Rozelle, Juan Flores Salinas.

India. Pooran Chandra, B. N. Chattopadhyay, S. R. Sen, V. M. Tandon, W. Garth Thorburn.

Java. A. T. Birowo, Soewondo Kartokeosoemo.

Malaysia. Karl Brandt, Chew Hong Jung, J. Norman Efferson, Mohamed Bin Hassan, W. T. Phillips, W. E. Taft.

Portugal. Richard T. McDonnell, Jose Maria Queiroga, John H. Sanders, João da Silveira, Carlos A. Vieira.

Taiwan. Chang Han Chang, Chiang Ching Che, Chiang Moo Ken, Thomas Hamby, Hong Tsing Piao, Hong Tsing Tien, Jiang Mu Gen, Lin Shih Tung, Tsai Wen Hai, Tsai Yao Kun, Tsai Yu Geng.

Union of Soviet Socialist Republics. V. I. Nazarenko, Nils Westermarck.

United States. John Kadlec, Dorothy and Robert Mills.

Upper Volta. Agoudiha Florent, Adama Balima, Joan Brakke, Ronald Cantrell, Sandy and Mahlon Lang, Ouedraogo Boureima, Ouedraogo Seydou.